SHORTCUTS TO SUCCESS

WITHDRAWN FROM
BROMLEY LIBRARIES

BCS, THE CHARTERED INSTITUTE FOR IT

Our mission as BCS, The Chartered Institute for IT, is to enable the information society. We promote wider social and economic progress through the advancement of information technology science and practice. We bring together industry, academics, practitioners and government to share knowledge, promote new thinking, inform the design of new curricula, shape public policy and inform the public.

Our vision is to be a world-class organisation for IT. Our 70,000 strong membership includes practitioners, businesses, academics and students in the UK and internationally. We deliver a range of professional development tools for practitioners and employees. A leading IT qualification body, we offer a range of widely recognised qualifications.

Further Information
BCS, The Chartered Institute for IT,
First Floor, Block D,
North Star House, North Star Avenue,
Swindon, SN2 1FA, United Kingdom.
T +44 (0) 1793 417 424
F +44 (0) 1793 417 444
www.bcs.org/contact

SHORTCUTS TO SUCCESS
PROJECT MANAGEMENT IN THE REAL WORLD
Second edition

Elizabeth Harrin

© 2013 Elizabeth Harrin
1st Edition published as 'Project Management in the Real World.' 2007

All rights reserved. Apart from any fair dealing for the purposes of research or private study, or criticism or review, as permitted by the Copyright Designs and Patents Act 1988, no part of this publication may be reproduced, stored or transmitted in any form or by any means, except with the prior permission in writing of the publisher, or in the case of reprographic reproduction, in accordance with the terms of the licences issued by the Copyright Licensing Agency. Enquiries for permission to reproduce material outside those terms should, be directed to the publisher.

All trade marks, registered names etc. acknowledged in this publication are the property of their respective owners. BCS and the BCS logo are the registered trade marks of the British Computer Society, charity number 292786 (BCS).

Published by BCS Learning and Development Ltd, a wholly owned subsidiary of BCS, The Chartered Institute for IT, First Floor, Block D, North Star House, North Star Avenue, Swindon, SN2 1FA, UK.
www.bcs.org

ISBN: 978-1-78017-171-5
PDF ISBN: 978-1-78017-172-2
ePUB ISBN: 978-1-78017-173-9
Kindle ISBN: 978-1-78017-174-6

British Cataloguing in Publication Data.
A CIP catalogue record for this book is available at the British Library.

Disclaimer:
The views expressed in this book are of the author(s) and do not necessarily reflect the views of the Institute or BCS Learning and Development Ltd except where explicitly stated as such. Although every care has been taken by the authors and BCS Learning and Development Ltd in the preparation of the publication, no warranty is given by the authors or BCS Learning and Development Ltd as publisher as to the accuracy or completeness of the information contained within it and neither the authors nor BCS Learning and Development Ltd shall be responsible or liable for any loss or damage whatsoever arising by virtue of such information or any instructions or advice contained within this publication or by any of the aforementioned.

Typeset by Lapiz Digital Services, Chennai, India.
Printed at CPI Group (UK) Ltd, Croydon, CR0 4YY

Some short sections have previously appeared on ProjectManagement.com or GirlsGuideToPM.com and are reprinted with permission.

For my family

CONTENTS

List of figures and tables	x
Author	xi
Foreword	xii
Acknowledgements	xiv
Glossary	xv
Preface	xviii

SECTION 1: MANAGING PROJECT BUDGETS — 1
Introduction — 1
1. Create a realistic budget — 3
2. Calculate the true cost — 7
3. Track estimate to complete — 11
4. Agree a budget tolerance — 14
5. Have a contingency fund — 18
6. Gain buy-in for collective budget responsibility — 21
7. Agree who holds signing authority — 26
8. Arrange for a peer review — 28
9. Manage projects with no budget carefully — 32
10. Use timesheets for tracking time — 35
11. Budget for change management — 38
12. Understand the benefits — 40
Further reading for this section — 42

SECTION 2: MANAGING PROJECT SCOPE — 43
Introduction — 43
13. Keep it small — 45
14. Know where you fit — 48
15. Work out how to manage changes — 51
16. Include quality planning in scope — 55
17. Work out how to track benefits — 58
18. Eliminate ambiguity — 63
19. Use version control — 66
20. Put a post-project review in scope — 70
21. Identify risks upfront — 74
22. Manage risks — 77
23. Manage issues — 83
24. Document assumptions — 86
25. Involve users in scope definition — 89
26. Communicate and document changes — 93

27.	Plan for handover into production	97
28.	Actively manage requirements	100
	Further reading for this section	102

SECTION 3: MANAGING PROJECT TEAMS — 103
Introduction — 103

29.	Set the vision	105
30.	Know the culture of your team	108
31.	Agree who is going to sign off	111
32.	Don't forget the soft stuff	113
33.	Train your sponsor	116
34.	Carry out stakeholder analysis	120
35.	Present your stuff interestingly	125
36.	Organise a party	128
37.	Make friends with the PMO	130
38.	Be a leader	133
39.	Manage a matrix environment	135
	Further reading for this section	137

SECTION 4: MANAGING PROJECT PLANS — 139
Introduction — 139

40.	Keep up the momentum	141
41.	Plan first – set end date later	143
42.	Manage fixed date projects carefully	146
43.	Have short tasks	149
44.	Understand the critical path	152
45.	Baseline your schedule	155
46.	Make meetings productive	158
47.	Delegate sub-plans to workstream leaders	160
48.	Manage project dependencies	163
49.	Manage multiple projects at the same time	165
	Further reading for this section	168

SECTION 5: MANAGING YOURSELF — 169
Introduction — 169

50.	Get organised	171
51.	Don't do zombie project management	173
52.	Keep your records tidy	175
53.	Manage your personal brand	177
54.	Navigate office politics	181
55.	Know what's a showstopper	184
56.	Learn how to facilitate	187
57.	Get a mentor	191
58.	Do documentation	194
59.	Don't be afraid to suggest they pull the plug	198
60.	Archive effectively	200
	Further reading for this section	201
	Join the conversation	202
	Share these ideas	202

Appendix 1: Risk log 203
Appendix 2: Issue log 205
Appendix 3: Change log 207
Appendix 4: Foreword to the first edition 209

End notes/References 211
Index 218

LIST OF FIGURES AND TABLES

Figure 2.1	Types of project expenditure	9
Figure 3.1	Calculating estimate at completion	13
Figure 4.1	Time and budget tolerances for a hypothetical project	16
Figure II	The golden triangle of scope, resources and time	43
Figure 14.1	The strategy triangle	49
Figure 17.1	Benefits management process	61
Figure 21.1	Risk matrix	76
Figure 25.1	Mental model mismatch	92
Figure 26.1	The change communication wheel	96
Figure 27.1	Increasing operational team involvement	99
Figure 44.1	A critical path	153
Figure 47.1	Example project organisation structure	161
Figure 50.1	Urgent and important tasks	172
Table 17.1	Types of success criteria	60
Table 19.1	Version control table in a sample document	67
Table 19.2	Tips for version control	68
Table 20.1	PPR checklist: example questions to ask during the post-project review meeting	72
Table 22.1	Risk responses for negative risk	80
Table 22.2	Risk responses for positive risk	81
Table 34.1	Mapping stakeholder interest and influence	124
Table 44.1	Task list for project to start up a collection scheme for recycling glass	153
Table 48.1	Types of dependency	164
Table 58.1	Standard project documents	196
Table 58.2	Example of an approval area on a typical project document	197

AUTHOR

Elizabeth Harrin, MA, FAPM, MBCS is Head of Applications and Programme Management for Spire Healthcare. Elizabeth spent eight years working in financial services (including two based in Paris, France) before moving into healthcare. Elizabeth is a PRINCE2, MSP and P3O Practitioner, and holds the ITIL Foundation certificate.

Elizabeth is also director of The Otobos Group, a project communications consultancy which helps people tell the story of their projects more effectively. She is the author of *Social Media for Project Managers* and *Customer-Centric Project Management*. Her interest in the uses of social media and new technology has developed over the past few years and she writes the award-winning blog, A Girl's Guide to Project Management. She is widely published on project management topics and has contributed to *Project* magazine as well as the websites Projects@Work, ProjectManagement.com and PMTips.net

You can find Elizabeth online at www.GirlsGuidetoPM.com or on Twitter @pm4girls.

FOREWORD

In an age where technology plays a protagonist role in our lives, the need to find and retain excellent project managers becomes more and more prevalent. The biggest issue facing companies these days is project failure, and there is the expectation that project managers can do more with less.

As a Six Sigma Black Belt, Certified Project Management Professional, and Industrial Psychologist, I've found that irrespective of what books or methodologies you use, the secret to project success is a simple but complex formula:

Project Success = (Scope + Cost + Schedule) + Soft Skills

Scope is basically what you are expected to deliver by the time your project is done. It's important to have that clearly defined up front, otherwise scope creep can happen – you wind up taking on more than you are supposed to. As tenured as I am in this field, I sometimes fall into this trap in the spirit of managing relationships.

Not too long ago, I had an internal client who asked for a simple solution. It was pretty straightforward and because I wanted to increase my profile with this internal client, I took on this work without doing the proper documentation (e.g., Project Charter, Business Requirements Document, etc).

My team and I were able to meet the tight turnaround time and deliver the outlined scope. However, once this person had the product in hand, all of a sudden this individual was saying, 'can you add this.....can you change that....oh, we also need this'. What I thought was going to be a relationship enabler quickly became a sour disabler. Making sure you document and have full transparency on what you agree to deliver is critical not only to your relationship with your sponsor but also to your brand.

Although cost sounds straightforward, it can be very complex depending on the organisation you work for. Some organisations view costs as solely the resource cost to manage the project. Others count the total cost of ownership. The difference is in how the costs impact the actual profit and loss statement.

The other consideration when defining and managing costs is the return on investment. Irrespective of how your organisation tracks costs, it's critical for project managers to show some sort of output value such as cost avoidance or expense reduction.

Cost avoidance can easily be calculated, for example, by asking the client how many hours it takes him or her to do the work manually compared to with a technology solution. How you set up this metric is really up to you and your sponsor. Your goal for every project is to think of and calculate the return on investment.

Lastly, keeping to the schedule you outline is critical to how well you and the team are able to deliver the results defined in scope. Recently, many companies have been faced with regulatory driven timelines, which results in forced work acceleration and in post-implementation rework.

Keeping to a schedule provides clarity and transparency to all stakeholders about what outcomes are expected at a particular time on the project. What I've found is that if there aren't hard deadlines, project managers tend to move the deadlines just to make sure they don't appear as though they are delivering projects late.

Elizabeth has done a fantastic job in breaking down the concepts I've applied in a way that you can easily adopt for any project. As you progress in applying theory into application, keep in mind that these techniques and tools are meant to guide you through your process. Good luck and have fun.

Bernardo Tirado, PMP
Chief Executive Officer
The Project Box, LLC
www.theprojectbox.us

ACKNOWLEDGEMENTS

I am indebted to the many managers and companies who have generously given their time and offered their experience for these case studies, including Bernado Tirado who wrote the Foreword to this edition.

I am also grateful to Jutta Mackwell and Matthew Flynn at BCS for their patience and support throughout the process, along with the anonymous reviewer who provided comments on the preparation of this edition. The book has been extensively proof-read and reviewed, so any errors or omissions in it are strictly my own.

GLOSSARY

Actual cost of work performed (ACWP) The amount of money spent on the project activities up until a given date.

Assumptions Statements made during a project that are not based on known or certain facts.

Baseline A stake-in-the-sand view of a project schedule, budget or other moveable activity which provides a comparison of the actual situation against the expected situation.

Business-as-usual Day-to-day activity as distinct from project activity.

Change control The process of managing change in a controlled way.

Change management See change control.

Contingency Provision made within the project planning stages to allow for unforeseen circumstances; usually built into the budget or schedule.

Critical path The longest route through a project plan; collective name for the group of tasks that must be completed on time in order for the project to deliver to the planned end date.

Critical path analysis The process of establishing the critical path; can include drawing out the critical path diagrammatically.

Deliverable Something tangible delivered as a result of the project.

Dependency A relationship that links the order in which activities are carried out; Task B is said to be dependent on Task A if the start or finish date of Task A must be reached before Task B can start.

Earned value analysis (EVA) A method to establish the budget and schedule position of a project based on resource planning.

Estimate at completion (EAC) The total budget required to finish the project, calculated by adding together estimate to complete and expenditure incurred to date.

Estimate to complete (ETC) The budget required to finish the project calculated from a given date to the project end.

Ice-breaker An activity or short game used to introduce team members to one another; used in workshops, long meetings or at the beginning of projects.

Issue A risk that has actually occurred or another known circumstance that may impact the project's outcomes.

Issue log A document listing all the issues that are impacting the project; updated with the activities required to actively manage and resolve each issue.

Issue register See issue log.

Management reserve See contingency.

Milestone A date when a particular chunk of work is due to be completed.

Network diagram A visual representation of a project plan, showing the links between each task; used in critical path analysis.

Pilot phase/stage A project implementation in miniature to test and assess the impact of the deliverables before the project is fully rolled out.

Plan A document, or several documents, detailing exactly what the project needs to do in order to deliver the objectives; a practical analysis of what deliverable will be produced by whom and when.

Post-implementation review See post-project review.

Post-project review A meeting to evaluate the project's successes and challenges and record any learnings for future projects; a way of sharing corporate knowledge.

Programme A collection of projects with a common theme, sponsor or reporting process.

Project board See steering group.

Proof of concept A test of the project deliverables in a controlled environment; shorter and more lab-based than a pilot.

Requirements document A document that records all the things (requirements) the end user wants from the project; used as a basis for technical documentation.

Risk A statement of the possibility that something unforeseen will happen to the project that will have a negative or positive impact on the outcome.

Risk log A document listing all the risks that may impact the project; updated with the activities required to minimise each risk.

Risk register See risk log.

Risk response The approach to managing a risk; typically one of: avoidance, transference, reduction, acceptance.

Schedule A document listing all the tasks that need to be done to complete the project and the dependencies between them; the project calendar.

Scope statement A description of what is included in the project and what is not; covers deliverables but also groups of people impacted and the reach of the intended activity.

Sponsor The senior manager who heads up the project; the person who champions the work and to whom the project manager reports with project progress.

Stakeholder analysis An exercise to determine the interest and influence of stakeholders to establish their support for the project and what can be done to influence their position.

Stakeholder mapping See stakeholder analysis.

Stakeholders Those people who have an impact on, or who are interested by, the project.

Steering committee See steering group.

Steering group A group made up of the project sponsor, project manager and one or two other key stakeholders; this group is responsible for decision making.

Success criteria The standards by which the project will be judged at the end to decide whether or not it has been successful in the eyes of the stakeholders.

Test scripts Documents explaining the step-by-step method required to test a deliverable; given to testers to ensure testing is done in a methodical way.

Wiki A collection of web pages acting as a hyperlinked knowledge repository and data set.

Workstream Part of the project that can be managed as a discrete chunk; led by a workstream leader.

PREFACE

When I was asked to put together a second edition of *Project Management in the Real World*, I wondered how much of the wisdom I had gathered for the first edition was out of date. Surely the shortcuts hadn't changed that much in six years – good project management practice was still good project management practice. I was wrong.

In the last few years, lots of things have changed. A global recession forced businesses to rethink how they deliver more projects with fewer resources. Keeping ahead of the curve became more important. The job of a project manager morphed from someone who was paid to get things done to someone who contributed effectively to business strategy through delivery. Project Management Offices became a more mature part of many businesses, helping to standardise and improve practices, and stop the loss of organisational knowledge. The role of social media tools for branding and communication grew beyond my expectations – and I was an early adopter of many of them.

Consequently, it has been possible to fully revise and update this edition. *Shortcuts to Success: Project Management in the Real World* won't teach you how to be a project manager. It's not going to show you how to set up your first project, walk you through it and see you out the other end with all the benefits realised. There are plenty of other project management books that follow the project life cycle with chapters on project definition, initiation, execution, closure and so on. This book is different.

It's for people who already know that a project has a beginning, a middle and an end and want to take project management further. It's for people who know the theory and feel there must be an easier way to get things done. It's over 250 years of project management experience distilled into 241 pages so you can see how other people run their projects outside the management texts and research papers: how projects get done in the real world.

The book is organised into five sections: managing the project budgets, scope, teams, plans and yourself as project manager. Wherever you are in your project life cycle you should be able to easily find information relevant to the particular situation you find yourself in.

Each of the five sections is divided into short chapters which explore discrete elements of the business of project management. Each chapter includes an anecdote from a manager who has been there and done it or a case study from a project with a valuable lesson to be learnt. Some names and project settings have been changed or disguised at the request of interviewees, but many of them have given permission for me to share their details and those of their projects. In addition, each chapter covers one learning point which you can put into practice immediately. The idea is that from reading about other people's experiences and a little bit of theory you will understand both why and how things can be done. Think of the book as your personal mentor, and an opportunity to learn from others.

Dip into the chapters at random and pick a section or make your way methodically through the section most relevant to where you are in your project at the moment. If a topic particularly grabs you, flick through the further reading suggestions and references to find ways to take it further.

Throughout the book you will see icons in the margins to guide you to important information in the text. Here's the key:

HINT

A hint or tip to help you apply the knowledge in the chapter.

ANECDOTE

An anecdote or case study: real-life experiences from project managers who have been there.

GOLDEN RULE

The golden rule to remember, even if you don't remember anything else about the chapter.

DEFINITON

A definition of a project management term or principle.

DANGER

Potential trouble spots or project management pitfalls.

The chapters cover the elements that I feel are most relevant to modern project management but are frequently overlooked. It has not been possible to include everything that I wanted to, and I'm sure you'll have a favourite hint, tip or memory that you believe other project managers could learn from. Please email me with your ideas for another volume at elizabeth@otobosgroup.com.

Elizabeth Harrin
London, January 2013

SECTION 1:
MANAGING PROJECT BUDGETS

INTRODUCTION

> *Know that with a farm, as with a man, however productive it may be, if it has the spending habit, not much will be left over.*
>
> Marcus Porcius Cato (234–149 BC), *De Agricultura*

More than one-third of projects have a budget of over £1 million so knowing how to handle the finances is an essential part of a project manager's repertoire. The initial budget is often just a starting point. An incredible 56 per cent of projects are affected by budget changes and that's not just a one-off financial revision. The average project, if there is such a thing, has its budget revised 3.4 times.[1]

Keeping on top of all this is not always easy, and it is made harder by the fact that project managers themselves don't always get control over the money. If that's the case, why should you care about the numbers? The answer depends on where you think a project manager's role ends. If you believe that your job is to deliver the project according to the scope and quality criteria set out by the sponsor, then it doesn't matter about tracking hours of effort or money spent. However, the project manager's role should cover far more than that. Your role is to deliver a project that is fit for purpose and adds some value to the organisation. Whatever you are working on should have a benefit, even if they are not financial benefits. There should be a purpose to what you are doing – someone who cares about the outcome enough to sponsor the project, and a business case that justifies why you and your organisation are bothering to work on this project at all. And that requires you to know a little about the finances of the project.

This section covers how to manage project variables over which you do not necessarily have authority, how to find out who has that authority, and how to manage the relationship with the budget holder. Many projects do not appear to have budgets at all and Chapter 9 looks at working effectively in that environment. This section also looks at reporting, tolerances and contingency.

1 CREATE A REALISTIC BUDGET

Even the smallest project will have overheads: your time as the project manager as a minimum. Nearly all projects will have more than that, so part of your role in setting up the project is to define and propose a budget for the work and get that approved.

PLANNING REALISTICALLY

Established in 1943, Hanford, a nuclear processing plant, produced plutonium for the world's first nuclear device. The facility, which lies along the Columbia River in Washington State, is now home to one of America's largest nuclear waste storage plants run by the United States Department of Energy (DOE). There are 177 waste tanks on site, storing about 56 million gallons of high-level radioactive waste underground – that's equivalent to an area the size of a football field over 150 feet deep.

The DOE launched an 11-year project in 2000 to build facilities at Hanford to treat and prepare the waste for disposal. Around the same time, the DOE launched a project management initiative designed to counteract the department's poor record of inadequate management of contractors. The initiative recommended that contingency funding be built into a project budget according to the project's degree of risk. Unfortunately at the time of signing a contract with the construction company in December 2000, the project management initiative had not been fully implemented. When an internal DOE assessment was carried out, it became clear that the department had signed a contract with a flaw: the cost baseline of $3.97 billion was so low that the project had only a 50 per cent chance of delivering against it.

The DOE took steps to address the gap in April 2003 and revised the cost baseline to include a $550 million contingency budget. They also set up a governance panel consisting of both DOE and contractor personnel to manage the additional funding and to monitor spending. The aim of the contingency budget was to counter unforeseen cost increases across the life of the project. The team also allocated an additional $100 million to be used to mitigate unforeseen technical and management risks.[2]

An audit in March 2005[3] highlighted that project reports were still showing that the clean-up work was on target to meet the approved baseline of $5.78 billion. However, by 2006 the construction project team was forecasting a final budget of $12.3 billion. The massive increase in cost was due to contractor and management performance problems, changes in scope and technical problems. As a result, the timescales had also slipped. The team had initially planned for the construction

> to be complete and for the treatment of waste to start in 2011. This has now been pushed out to 2019[4].
>
> The Government Accountability Office, which audits major public sector projects in the United States, reported in 2009 that the DOE's estimates of how much it will cost to complete this project 'are not credible or complete'. They have also criticised how estimates have grown each time the work has been re-estimated. At the moment there is no way of knowing exactly how much this project will cost – or if it will finish successfully at all.

You might not be decommissioning thousands of tons of nuclear waste, but you can learn from the need to create a realistic budget. You can work out how much money you will be spending based on what you know needs to be done, just as you work out how much time the project will take based on the same information. Think of the budget as a shopping list of all the things you need to buy to make sure the project gets completed. Just like a trip to the supermarket, you might not end up spending exactly what you expected but at least the list gives you a reasonably accurate starting point. 'When planning, assume your budget will not be increased or decreased during the project,' writes George Doss in the *IS Project Management Handbook*. 'Budget changes…are adjusted through negotiations with the project sponsor based on circumstances at the time.'[5]

There are five steps to creating a project budget:

1. Identify the resources required for the project.
2. Estimate the cost for each of those resources.
3. Document the costs and calculate the overall figure.
4. Submit the budget to your steering committee or sponsor for approval.
5. Find out your budget code.

Let's take each of those steps in turn:

1 IDENTIFY THE RESOURCES REQUIRED FOR THE PROJECT

Review the schedule, project initiation document and any other documents you have to identify the activities that need to be completed. Draw on your stakeholders and project team to brainstorm anything else that might be required (like travel, accommodation, couriers, equipment and so on). Will your project have to pick up the costs incurred by other areas of the business that are impacted by the work you are doing? Ask other managers who have done similar projects to validate your list.

2 ESTIMATE THE COST FOR EACH OF THOSE RESOURCES

Every step, every task of the project will have associated costs. Projects that do not have full-time staff may avoid paying for the entire salary of anyone working on it, so ask the finance department if there is a list of standard chargeable rates per 'type' of employee. For example, your project might have to pay £1,000 per day for an expert manager, but

£650 per day for a junior marketing executive. Some of these costs may be just 'wooden dollars' – especially for internal resources. They are simply figures you plug into the business case but in reality money never changes hands. Check out your company's rules for charging for project team members' time and also check with each department head about their expectations. For example, if they are loaning you a person for the team, they may expect the project to fund a temporary resource to back-fill that person's day job.

A NOTE ON ESTIMATING

Given the flexible nature of budgets, and projects in general, it's very hard to pin down costs to an exact figure at an early stage of the project. And it's not a good idea either, unless you are absolutely 100 per cent sure that your estimate is spot on and will not change.

At this stage, present your estimates as a range rather than a fixed sum. This means that your overall project budget, once you have added up all your estimates, will be between £x and £y. It is this range that you present to your project steering group and sponsor.

Presenting a range gives you a little more flexibility later on. It also offers you the chance to start managing the expectations of your key stakeholders now – they will have to come to terms with vagaries and changes as the project progresses so now is a good time to start explaining the nature of project management.

3 DOCUMENT THE COSTS AND CALCULATE THE OVERALL ESTIMATE

Companies that carry out a lot of projects will probably have a standard template for submitting a budget, so find out if there is a form that already exists. Create your own in the absence of anything standard, using a method that suits you. Spreadsheets are the most effective way of recording and managing costs. The advantage with an electronic budget spreadsheet over using a word-processing package or a paper system is that you can include formulae to ensure that summary figures and column totals update automatically, reducing the risk of manual error and saving time. Software like Google Docs (if this is authorised for use by your company) allows you to share the spreadsheet in real time with your project team and stakeholders, wherever they may be, and have multiple people update it (although you may not want this, of course). Group similar costs together so you have sub-totals as well as an overall total and include a line of contingency for risk management. Compare your budget range to any amount given to you by the project sponsor and see below for what to do if the figures don't match.

4 SUBMIT THE BUDGET TO YOUR STEERING COMMITTEE OR SPONSOR FOR APPROVAL

Once you have your budget written down, it needs to be approved before the project can continue. Your sponsor or steering committee are the first point of approval. They will advise you on whether the budget needs another level of approval from finance, a central planning committee, an IT authorisation forum or another group, depending on where the funds are actually coming from.

More often than not, you'll be asked to kick off the project without budget authorisation. In the real world, there are deadlines to meet that won't wait just because the budget committee only meets on the last Tuesday of the month. If you're asked to start work without the relevant approvals – get on with it! But make sure you have something in writing to cover yourself against any expenditure incurred during the time you're working without an approved budget.

5 FIND OUT YOUR BUDGET CODE

Assuming all goes well, the budget will be approved and you will be given the go ahead to spend the money required. Any expenditure needs to be tracked back to the project so the budget holder can keep an eye on what is being spent. The project might be allocated its own 'pot' of money, ring-fenced from other budgets, in which case you will probably have a cost centre code of your own. Alternatively the project might be allocated a portion of the budget for a particular department. If this is the case ask your sponsor how they want you to identify project spending. A non-committal answer means you will have to invent your own code, perhaps the project number or a shortened version of its name. When you sign to approve an invoice or raise a purchase order, use the code to ensure the expenditure can be tracked back to the project: make certain that anyone else who has the authority to use the budget does this as well.

WHAT IF MY SPONSOR ALREADY HAS A BUDGET IN MIND?

Just because this is a sensible five-step approach, allowing you to analyse the work involved and cost it accurately, does not mean that it is followed by all project sponsors. For many reasons you could find yourself working on a project where the sponsor already has a set figure in mind. Some sponsors will knock off 10 per cent from your total because they believe the numbers are padded. Others may be compelled to halve the budget because someone higher up the chain expects cuts across the board.

If you put your mind to it, you can complete any project to a specified budget: at a hidden cost. Corners will need to be cut, quality might suffer and the customers may not get everything they thought they would. Present your steering group with a couple of options for reducing your proposed budget to their predefined figure, making the trade-off between quality, time, scope and cost. They may still tell you that it's their budget you need to follow, but at least you have explained the risks of delivering to a certain, abstract, budget figure and you have your planning documentation to back up your arguments.

To create a realistic budget, base your predicted expenditure on your project planning documentation and get the budget approved as quickly as possible to prevent any delay in starting work.

2 CALCULATE THE TRUE COST

The cost of your project is probably not as transparent as you originally thought. Digging into the detail will help you really understand how much the project will cost and help you avoid any nasty surprises later.

THE COST OF CARS

'One early project I worked on required the consolidation of nine separate service desks into a central service desk,' says Alison Marshall, a US project manager with 17 years' experience. 'The existing service desks were at different physical locations. During a team meeting someone asked where they should park when they began working on-site.'

This started a discussion about the cost of parking – and these costs had not been added to the initial budget. Staff members had free parking at their current work locations but the new site charged $120 per month for parking. Parking for over 60 people soon added a massive $90,000 cost to undertaking the project.

'This indirect cost was significant,' Alison explains, 'whether the project paid for the parking, the organisation supplemented the difference via salary or absorbed the turnover cost of losing staff unwilling to pay the new fee.'

The organisation decided to supplement the fee by providing salary increases to the affected staff. 'As project sponsors use final reported costs as a benchmark for similar future work it was important to review the original budget estimate against all true costs associated with the project,' Alison explains. 'The true cost variance from estimate when the parking fees were added was 3.75 per cent. In my experience a key to deriving true project cost is to place an emphasis on identifying and capturing indirect costs. Don't be afraid to ask as many questions of as many people as you can. You can't calculate what you don't know.'

When finalising a project budget Alison combines the direct and indirect costs for owning an asset, project or system to provide the total cost of ownership with the cost of carrying out the associated project activities. 'In an attempt to gather costs for the full project lifecycle I use a collaborative approach,' she says. 'I request feedback from the project team members regarding which activities and items are needed to execute the project. Once team members feel that they have had their individual

> input I bring the team back together for brainstorming. You will be surprised by what comes out of that type of team session. Something you perceived as a small issue or one you may not have thought of on your own can have significant cost implications.'

> Shim and Siegel define a budget as 'the formal expression of plans, goals, and objectives of management that covers all aspects of operations for a designated time period'.[6]

There are two important elements here. The first is 'all aspects'. Figure 2.1 shows the two categories of expenditure you should consider in detail:

1. project management costs: the costs of doing the business of project management, and;
2. project deliverable costs: expenditure directly related to what the project is going to deliver.

Using these two categories will help clarify your thinking when you are analysing your budget to guarantee that it includes 'all aspects'.

You might be working on a smaller project and not trying to manage the consolidation of nine service desks and the knock on implications of 60 people's car parking fees, but the lessons from Alison's experience are worth implementing nevertheless. When you are trying to calculate the total cost of your project, brainstorm all the things that will cost money in both categories. Do this with your team so that collectively you have the best possible chance of identifying everything. Include paying for your team members (project management costs), buying software or consultancy and funding anything that will change as a result of your work like new stationery or user guides (project deliverable costs). Maybe you will have to provide training courses (project management costs which will include the costs of a trainer, room hire, refreshments, delegate transport and accommodation), host large meetings off-site (project management costs) or pay for documents to be translated (project deliverable costs).

Once you have a comprehensive list, add to it all the things that you believe will not cost anything: the business users' time for testing, your time and so on. This gives you a documented list of 'free' things. These form project assumptions and will be validated as the project progresses. Your sponsor should verify this list, which can be included in the project initiation document. If at any point you find that you were wrong and that you do have to pay for items you believed were free, you can explain to your sponsor that the budget increase is due to these assumptions being incorrect. We'll see more about assumptions in Chapter 24.

Once you've worked out all the elements and likely costs, decide on how to report your budget. If your sponsor is only interested in cash out the door, it will not be necessary to report how many days the 'free' human resources have spent working on the project.

The second important phrase in Shim and Siegel's definition is 'for a designated time period'. The project does not necessarily end as soon as you have delivered whatever it was you set out to deliver. There could be (and are likely to be) costs incurred in the final stages; the 'grave' part of 'cradle to grave'. The budget should include adequate provision for any end-of-project expenditure. That means, for example, the charges for decommissioning any now-defunct system, product or literature, retraining for staff who now don't have the right skills and generally making sure the old status quo is not someone else's financial headache.

Michael Cavanagh, in his book *Second Order Project Management*, argues that you don't know the full cost of a project until it has been delivered. '[P]ost-delivery costs including fault correction, maintenance, support and disposal are all subject to the vagaries of implementation in the real world and should be addressed and included in the estimate process,' he writes.[7] It is very difficult to estimate these elements, as at the beginning of the project it's almost impossible to know what they could be. You can use the budgets of previous projects as a guide, and conversations with subject matter experts as well to see if you can come up with some appropriate estimates for the post-delivery phase.

Figure 2.1: Types of project expenditure[8]

Project Management Costs   Project Deliverable Costs

The post-delivery phase signifies the end of your project life cycle. If the whole life cycle stretches over 12 months this can have another important impact on budget management. If your project stretches over two financial years you will have to apportion your budget appropriately and might have to navigate your way through the maze of year-end accounting and the accrual process. Get some advice from an old hand if you're facing doing this for the first time as the rules differ from one organisation to another – although it might take you a little while to find someone who can explain them clearly!

Don't guess what your budget is supposed to pay for. Do your own research and work with your team to fully understand all the explicit and hidden charges to help you control costs more accurately over your project's life cycle.

3 TRACK ESTIMATE TO COMPLETE

It is easy to track how much of your precious project budget you have already spent (assuming you keep copies of invoices and timesheets). It is also easy to assume that you will simply use up the rest of the available money evenly between now and the end of the scheduled work. However, this approach gives you a false impression of how you will spend the rest of your budget.

To give you an example, you could be 50 per cent of the way through the project and have spent 50 per cent of the budget. However, if a big purchase has not yet happened, such as buying and installing servers for a production software environment, having only half the budget left could be a sign of a trend towards overspending. Project costs are rarely distributed evenly throughout a project: some projects spend very little early on and then incur all the costs in the last few weeks.

'Estimate to complete' (ETC) is the amount of money that you predict will be required to finish the project. Tracking ETC gives you an accurate view of the projected budget you need to get everything done.

REAL-TIME BUDGETING

'The most important things I reported on projects were budget to date, actual to date and estimate to complete,' says Lonnie Pacelli, who worked at Accenture before setting up his own consultancy. His 20 years of experience at delivering projects and his previous role as Microsoft's director of corporate procurement overseeing $6 billion of expenditure have given him a clear insight into how to handle project budgets.

'One of the most overlooked components of budget management is a realistic estimate to complete,' he continues. 'Too many times a project manager will just subtract their actual to date figure from the total budget to calculate the estimate to complete.' Lonnie believes this gives an unrealistic view of the money needed to finish the work. He suggests a better route is to determine the estimate to complete based upon the tasks the project manager knows are still to do, and not on a vague hope that the project will come in on budget.

'A realistic view of estimate to complete is a major sniper in the weeds on projects. Too many project managers assume that everything will go perfectly for the remainder of their project even if the project to date has been difficult,' adds the Washington-based president of Leading on the Edge International, a management consulting and self-study leadership education firm. 'Don't just say "I have $1 million total budget, have spent $600 thousand so my estimate to complete is $400 thousand." Look at the remaining amount of work left to complete and realistically cost out the work.'

This approach is more robust and it will also allow you to highlight potential budget issues early. 'I reported budget status to the executive sponsor, steering committee and also the project team,' Lonnie says. His advice for project managers who find themselves reporting that their projects will go over budget on completion is to use a realistic ETC figure and ask for more money just once. 'Going back to the well more than once erodes management's confidence in the project manager and creates doubt as to whether or not there is another surprise waiting around the corner,' he explains. He cautions against holding back the information about the cost rise, and is clear that raising the issue early is a safer tactic. 'Don't assume some wonderful thing is going to happen which will cure all of your budget ills,' he says. 'Surprises are for birthdays, not for project budget management.'

Calculating the estimate to complete for your project is really straightforward. You already know how much of the work is done, so you know how much is left to do, based on the latest version of your plan. You can work out how much the still-to-do work will cost, based on your budget assumptions. That figure is your estimate to complete. It is more useful to express this figure as a financial amount rather than a number of hours or days of effort to be sure the ETC takes non-resource costs into account, such as buying equipment or leasing a property. Also bear in mind that you may not have all the invoices from third parties. Their charging patterns may mean they invoice a month in arrears, after the work is done. You could find yourself in a situation where the work is complete but not yet charged for. Just ask them for the relevant figures if you're not sure.

The ETC plus the figure you have already spent represents the budget you expect to have spent at the end of the project.

The budget already spent is known as actual cost of work performed (ACWP). Your estimate of the total amount spent on the day you close the project is known as estimate at completion (EAC) and this is the figure that will interest your sponsor. This calculation is summarised in Figure 3.1. Keeping track of EAC is a simple way to predict budget overspend. It provides an early-warning mechanism and allows you to plan how to tackle an increasing budget. In fact if you are working on a technology project it is highly likely that your monitoring will show escalating costs. Between 30 and 40 per cent of IT projects fail to stick to their original budget.[9] Studies show it is worse in other industries: 90 per cent of transport infrastructure projects overspend.[10]

The reasons for overspending are many and varied but the costs to be most aware of are those associated with your human resources. In fact, research shows that we are getting worse at managing the resources associated to the project: a study published in the *Project Management Journal* in 2010[11] showed that 26 per cent of people thought that they had enough resources allocated to their project in 2000 but only 16 per cent of people thought the same in 2008. The same study also showed that good cooperation with other teams and good work processes has a direct relationship to improving the chance that the project will complete on budget, so don't underestimate the role that people, following robust processes, can play in making sure you hit your budget. Fortunately, using ETC and EAC to track expenditure on your project can help you identify any trends towards extra costs at an early stage.

Figure 3.1 Calculating estimate at completion

$$\text{ETC} + \text{ACWP} = \text{EAC}$$

ETC, ACWP and EAC are components of a larger financial tracking method called earned value analysis (EVA). EVA is a tool that shows whether you are over or under budget, behind or ahead of schedule, at any given moment in the project.

There is not the space here to explain EVA in its entirety. If you are just starting out in project management and have a relatively small budget, calculating ETC and EAC are a good solid start. As EVA takes time and effort to do properly it adds limited value to small projects. With a larger project you may find the EVA method useful to help you understand where you are. *Project Management for Dummies* has an excellent appendix describing EVA and including worked examples so if you are interested in taking these measurements further, try that as a basic introduction.

Remember: any financial or mathematical calculation can only offer a numerical representation of a project's situation with regard to the overall costs and schedule. No figures like this will ever give you a narrative explanation of why your project has ended up here or why it appears to be going off the rails – you will have to work out the 'story' for yourself.

Tracking ETC and EAC will give you an early warning of possible project overspend and useful real-time information to report to your sponsor, but you will need to add the explanation of why the numbers look like they do for yourself.

4 AGREE A BUDGET TOLERANCE

A budget tolerance is a range within which you can spend without having to report back to your sponsor or ask for more money.

Budget tolerance is particularly useful at the end of a project as you near the delivery date. If you have a budget of £80,000 with a tolerance of 10 per cent and you complete the project for £85,000 you have still delivered within the parameters set by your sponsor. A budget tolerance of 10 per cent means you can deliver the project 10 per cent over cost without having to get special permission to do so.

NOT A PENNY MORE...

Peter McDonald, an engineer working in Wales, thinks back to his first project: 'It was quite small actually,' he explains. 'I was just starting out in project management and was working in a team improving the process for getting car parts off a distribution line more quickly. The project budget for non-resource spend was small and as no one else wanted to do it, I got put in charge of monitoring the expenditure.' The project manager delegated the responsibility for tracking the budget and ensuring the team did not spend more than had been agreed to Peter. 'I was really nervous and I watched every penny,' he adds. 'I suppose it was about £30,000, which considering what I manage now really wasn't that much, but at the time on my just-out-of-university salary it was massive.'

Peter's team had eight months to analyse the existing process, come up with a new one and implement the changes successfully. The analysis went well and within three months the team had got agreement from the factory management to implement their new process. There was no budget for buying new machines so the changes were subtle but effective. 'We ended up by streamlining the process in the warehouse,' Peter says. 'We couldn't make changes to the actual manufacturing part of the process as it was prohibitively expensive, but we cut out some of the admin steps.'

It was re-engineering the paperwork that used up most of the budget. The project manager consulted with Peter and purchased a system for hand-held scanning machines to remove the need for manual checking when boxes of the car parts were ready to be shipped out. 'The technology seems antiquated now, but it was revolutionary for us,' Peter says. 'But the problem came when the invoice dropped on to my desk.' He had forgotten to add the cost of delivery charges and the three-year warranty the company had purchased. With those additional amounts the budget was now running at three per cent over. Peter started to worry. 'There was no way I could pull the budget back in line, especially as I wasn't the project manager,' he explains. 'So I had to confess.'

Peter took the project manager to one side and informed him of the mistake. 'He asked me if we were still on track to deliver everything else within budget, and I said yes. Then he told me not to worry as he had agreed a five per cent tolerance with the sponsor!' Peter was relieved but annoyed. 'I should have been told that at the beginning when I was given the responsibility, but I didn't ask either,' he says. 'Since then I've made sure I know what the tolerance levels are for my projects so I'm aware if there is some degree of flexibility.'

At the beginning of the project discuss a budget tolerance with your sponsor. It is a way of minimising effort for them as you will not be bothering them with frequent change requests for tiny budget increases. Agree an appropriate tolerance and write it into the project documents. What is appropriate will depend on the size of the project, the size of the organisation and its maturity with regard to projects. The tolerance will not be 'used' until the end of the project but it will help you monitor performance and track how you are doing compared to your initial estimates. As soon as the project looks like it will fail to deliver inside the tolerances, you know you have a problem to address. Tolerances can be used like early warning systems: they give you a little bit of leeway but enable you to quickly tell how far you are from your targets if the project begins to stray off course.

HOW IS CONTINGENCY DIFFERENT FROM TOLERANCE?

A contingency fund is an amount of money set aside for project emergencies. It is a project's overdraft. The project manager needs permission and a good reason to spend it, but it is assumed that it will be used at some point or other through the project. Contingency can be for any amount, sometimes even 50 per cent of the original project budget. The project manager calculates an appropriate amount based on the project's risk factors and negotiates the final allocation with the sponsor.

Budget tolerance is the amount by which the project can be delivered over (or under) budget without anyone being concerned. It's usually a small amount represented as a percentage. Tolerance is either calculated as a straight percentage of the core budget estimate or as a percentage of the core estimate plus the contingency fund. As you should assume the contingency will be spent, it's better to agree a tolerance based on the latter.

The amount of tolerance is set by the sponsor or main budget holder, based on your recommendation. It's an acceptance of the fact that you might need a little extra and that in the grand scheme of things, the overall company budgets can handle a little flexibility.

This type of tolerance relies on your sponsor agreeing to a degree of flexibility within the project budget. But what happens if they say no? If you are not allowed an explicit tolerance then the pressure is on to deliver on budget. A contingency fund becomes even more useful. But what if you can't get your sponsor to agree to one of those either? Consider padding your budget estimates a little so you give yourself a cushion of implicit contingency. It's sneaky but it will give you more flexibility with the finances later if you can get away with it.

Budget and time tolerances are often set together as part of the same conversation with a sponsor, which means they can be plotted graphically as in Figure 4.1. This graph shows that the sponsor is happy for Project Whirlwind to finish between mid-September and mid-October and cost between £71,250 and £78,750 although the target is to finish at the end of September and spend £75,000. The sponsor and project manager have agreed a budget tolerance of +/– five per cent.

Figure 4.1 Time and budget tolerances for a hypothetical project

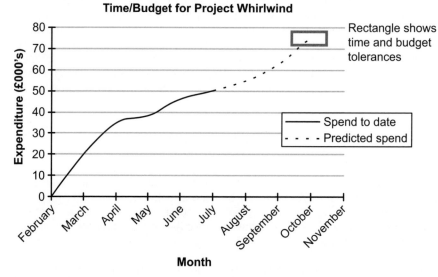

If your project is part of a programme, you may find that the programme overall has tolerance levels set. Your project will be expected to operate within these. Let your programme manager know if you are no longer on track to hit your budget targets exactly – he or she will need to balance the needs of other projects in the programme and may need to adjust the forecasts at the programme level to account for the change in your project.

Minus tolerances are important too. If you bring a £75,000-project in on time and to the required specification but only spend £20,000 then someone will start asking questions. Tying money up in project budgets that could be put to better use elsewhere is not good business practice and will cause significant issues in small organisations. If you believe your budget will not be spent within the tolerance levels, study why, double-check and then raise a formal change to make your sponsor aware of the issue and adjust the budget according to your new calculations.

Agree a budget tolerance early on with your sponsor, even if you won't come to use it until later in the project.

5   HAVE A CONTINGENCY FUND

A contingency fund is money set aside at the start of a project to be used in case of need, for example to offset unforeseen increases in costs. The amount of this ring-fenced budget depends on the level of risk the project faces and also on the overall project budget itself. Contingency funds are sometimes called management reserves.

PLANNING CONTINGENCY BASED ON RISK

'The contingency budget is the budget set aside to deal with identified and unidentified risks,' explains Graham Inglis, a technical project manager with 20 years' experience who is now based in Paris, France. 'After quantitative risk analysis, a budget for the identified risks should be known. In addition to this it may be wise to set aside a provision for unidentified risks, especially if the budget for the identified risks seems small. Some organisations have rules of thumb which set the contingency budget at say 10 per cent or 15 per cent, but in general it's sensible to modify this in light of the risk analysis.'

Graham was able to put this into practice on an electronics development project where the quality of certain design inputs was unknown. The team calculated the risk budget based on the cost of having to re-engineer these inputs and the impacts on staffing and on the schedule. Then they added a standard contingency budget for the later project steps based on previous experience on similar projects. 'After checking the margin impacts of the worst case scenario, the contingency budget was set based on the most likely case scenario,' says Graham.

As it turned out, the team did have to do some re-engineering. 'Part of the sum allocated to this was used,' Graham says. 'The remainder was retained in the general contingency budget for possible later use. Much later in the project, a problem was found which required a small redesign charge and a large re-manufacturing charge. The contingency budget had to be used.' Even so, there were still some funds left at the end of the project and these were released to general funds at the end.

'It's important that the project manager can easily access the contingency budget,' advises Graham. In the case of his electronics project, use of the contingency budget was under the control of the project manager, but reported to the project board. 'If the approval process is too complicated or lengthy there's a real risk that the project manager will not initiate the corrective actions early enough to contain the overrun to a minimum.'

At the beginning of your project you will need to calculate the expected budget required in order to deliver the work. This budget figure is the first step to being able to work out a reasonable contingency. The project risk log will also be important, as your knowledge of the risks inherent in the delivery will inform your decision about how much contingency is required: the riskier the project, the higher the amount of contingency budget.

Calculating contingency is not really a science. Once you have a full understanding of the work the sponsor expects you to deliver you must take a best guess at the figure and ask them to approve it. Your company might use a formula for calculating contingency but as it depends so inherently on the risk factors for each individual project it is hard to give a one-size-fits-all equation. A project that has been run several times before, with experienced staff and solid estimates for both time and expenditure, will need little, if any, contingency. A project using new technologies or doing something that the company has never done before will require a large contingency fund to offset any unforeseen disasters.

> **'IT'S TOO HARD TO WORK OUT WHAT "REASONABLE" CONTINGENCY IS. I WON'T BOTHER.'**
>
> It is not a good tactic to avoid setting a contingency budget because it's too hard to work out what is reasonable. Add a contingency line to your budget of 10 per cent. This at least gives you a starting figure to begin negotiations with your sponsor and it will give you some leeway if it does turn out that your estimating has been a little wayward. The more experience you have with budgeting, and with projects, the easier it will be to predict what amount would make a reasonable contingency fund.

Whatever your chosen figure for the contingency budget, you will have to convince your sponsor you need this allocation. Explain that it will cover things like cancellation costs for training courses when your delegates are ill, unforeseen bills (small items soon mount up) or VAT on a supplier's quote that you took to be inclusive but was actually extra. Aside from these small expenditures that you may not have seen coming, it will also be used to deal with any plans to address risk factors. Remember that the contingency fund is there to help cover costs for work that is required to meet the original objectives. If the sponsor wants changes to the project scope, these should be costed and their impact analysed separately. The sponsor should find additional funds to pay for changes. For more on managing changes, turn to Chapter 15.

Sponsors aren't renowned for letting project managers have what they could see as a slush fund. It may help your case to explain to your sponsor that the money will be kept separate from the main budget. It will not be physically in a separate bank account but it is a separate line in your budget tracking. Also explain the situations in which you will call upon those funds and the approval process required to spend them. This process could be that you'll ask for their approval every time you dip into the fund, or that you'll control the fund as a separate budget from the main project budget. Fleming and Koppelman, in their book *Earned Value Project Management*, argue that the second approach is better, but if the project manager is in charge of the contingency fund, that this be put aside. If the contingency budget is left within the main project budget 'it likely will be consumed' they write.[12]

Once you've agreed the process of using the money with your sponsor, you can also explain that having the additional money set aside now will make dealing with any project changes or unforeseen events much less painful in the future – and if you do not use the budget it will always be there for another project.

Ruthanne Schulte from Welcom[13] outlines four steps to managing budget contingency:

1. Calculate the amount of contingency budget for each short phase or unit of work, not at a project level.
2. Hold the money separately and get management approval to move it across to the main budget.
3. Once management have approved the spending, increase your budget appropriately so you have an accurate idea of how much you have spent or are predicting to spend.
4. If contingency is not required, give it 'back' to the company so it is not included in any profit calculations at the end of the project.

The important point here is to make sure that you report the use of any contingency funds transparently. It will give your sponsor more confidence that you are acting in an accountable way and should also dissuade you from overspending irresponsibly.

A contingency budget allows you to react more quickly to any unforeseen events that plague your project. If you are able to convince your sponsor to let you have one, set the amount based on project risk and ensure you have a mechanism by which to authorise and monitor the use of the funds.

6 GAIN BUY-IN FOR COLLECTIVE BUDGET RESPONSIBILITY

A budget is the responsibility of everyone in the project team, whether or not the team members individually have anything to do with the administration or invoice handling on a day-to-day basis. Every member of the team incurs costs (for their time as a minimum) and therefore should appreciate their role in making sure the project stays within budget.

MANAGING THE MEADOW

Roffey Park Institute is a charitable trust which is internationally recognised for developing innovative learning approaches that enable individuals to achieve their full potential both at work and in their wider lives. Since it was founded in 1946 the training courses and research have shifted with the economy to concentrate less on the factory as a workplace towards the well-being of employees in businesses. The success of the Institute led its Board to consider expansion – a project they knew would be a huge undertaking. 'We set a project budget and as Roffey Park is a charitable non-profit making organisation without substantial free reserves there was no possibility that this budget could be exceeded,' explains Val Hammond, who was chief executive at the time.

Working with a professional project manager, the Institute appointed architects who worked closely with the initial project team to understand the requirements and constraints. The brief was complex: redevelop the site of the Roffey Park Institute to replace existing residential accommodation with double bedrooms of four star hotel standard, add a conference centre, provide additional dining and catering facilities and incorporate suitable landscaping to link and blend the new and old buildings (some going back to the 1850s) into a harmonious whole.

'The architects' design exceeded everyone's expectations in terms of meeting the brief and also took into account the environmental issues and the need to keep running costs and on-going maintenance as low as possible,' Val says.

But the project soon hit a major problem. Roffey Park is situated in the Sussex countryside in an Area of Outstanding Natural Beauty (AONB) – zones that are carefully regulated and subject to very rigorous planning requirements. 'AONBs also tend to arouse strong emotional responses with the community since, by definition, they are beautiful and rare places,' Val continues. 'In this case the pre-planning application consultation with local residents took considerably longer than foreseen.' In fact, the consultation process delayed the original opening of The Meadow complex by around two years.

'We have been supporting the local community, as well as delegates from around the world, for decades, so it was important to us to do the consultation and application properly,' Val says.

Once planning permission was achieved, a permanent project team was appointed and everyone was confident of their ability to build within the budget figure. The team produced a new schedule for the project and agreed the dates. However, the Meadow project hit another problem as soon as the detailed building drawings were available as the construction company quickly formed the view that it could not be built for the agreed budget. 'Prices had risen since the designs were developed and they estimated the project, as planned, would now exceed the budget by nearly half as much again,' Val explains. 'We were not able to increase the available funds, so this was a much more serious problem than the time delay.'

The project team presented many options to reduce the cost, including losing the conference room, and cutting the number of bedrooms, but the Institute was utterly committed to the flexibility the original design offered and the building in its entirety had become integral to the business plan. The Institute, renowned for its innovative approach to management research, applied some of those skills to come up with a new way of bringing the project back under budget.

Val continues: 'Diligent work and painstaking negotiation on the part of all involved brought the project back into line. No elements were cut entirely but some were reduced in scale. Elsewhere detailing was streamlined to simplify the construction but distinctive elements such as the "egg shaped pods" to be used as syndicate rooms, the environmentally sound construction and living roof over the conference centre were retained.'

In a project where there was no margin for error and very little in the way of a contingency budget, Val knew that keeping the budget on track for the duration was going to be a challenge. 'The principal actors: the Institute as the employer, the architect, the quantity surveyor, the construction company, the site and services engineers as well as the project manager accepted joint responsibility for the budget,' she says. 'We agreed that if any element went off track for any reason, there would be a collective effort to find a saving elsewhere, without compromising the quality of the building.' It was an approach that proved to be very successful. 'Sometimes it involved one discipline finding ways of offering more cost-effective solutions in their area to allow another to meet a necessary overspend that could not have been anticipated. On occasions, it meant the Institute had to compromise too.'

The project team worked extremely closely together with a high degree of trust and there was architectural support on site continuously to deal with design queries and develop cost-effective solutions. Working in this way, the whole team brought the new buildings to completion at the agreed budget and on time. The Meadow opened as expected and was awarded a commendation by The Civic Trust. 'Even more important is that the building fulfils its purpose, the different elements work effectively and the facilities are always in demand,' Val says. 'Maintaining a clear focus on the primary objectives – a quality building that fulfils its purpose – and respect for each and every member of the team helped enormously in creating a sense of shared ownership and pride in our achievement. This, in turn, enabled us to achieve the impossible: to bring a building project to completion on budget!'

Budgets are, by their nature, collaborative. One person is not the root of spending all the money, even if one person approves all the expenditure. Different team members have their part to play in recommending suppliers, keeping you informed about what needs to be purchased and so on. Encouraging everyone to work together to keep costs in check and to report accurately will generate a sense of collective responsibility for the budget within the team.

Present the overall budget to the team at the outset of the project. At this point a high-level overview will do, concentrating on the overall amount to be spent and how that will be broken down. Explain how the budget processes that affect them work: that you expect them to use a certain cost centre code for their timesheets, for example.

Sharing the project budget with the team has several positive effects. It:

- gives team members a sense of the scale of the work;
- helps them to feel part of the entire project instead of just their tasks;
- educates them about the costs of their work in relation to the project;
- encourages them to report accurately as they see the impact their data has on the overall budget;
- highlights to the team the speed with which money is being used on the project;
- raises awareness of the relative costs related to each task;
- promotes a sense of individual responsibility: many team members will not have been entrusted with this type of information before;
- promotes a sense of collective responsibility: team members can see the interdependencies on other departments at a budgetary level;
- presents an opportunity for the team to challenge the budget or to add or revise estimates, and;
- establishes you as a project manager willing to share information, setting a good example for your team.

The budget may be fixed but there might be some degree of flexibility about how it is spent. Having collective awareness and responsibility for the budget, even if you keep the authority for it, can prove a useful strategy when the finances are relatively flexible. Team members can really see how overspends in their area affect other departments.

Report the budget status at each team meeting: what is on track, which tasks risk going overspent and which look underspent. Present a recommendation for reallocating the budget and negotiate with the team until you reach a decision which works for everyone. Can a task be done by a less experienced (and therefore cheaper) resource? Can an expensive piece of equipment be replaced by something cheaper but of the same quality?

Reorganising the budget within the team is one approach, but as soon as the solution to your budget concerns is to change the quality, alter the way in which the project will deliver the required objectives or change the time frame you must get the sponsor's approval. Anything that fundamentally changes your time, scope, budget and quality 'contract' with the sponsor must be ratified by them and, if necessary, the steering group. J. Nevan Wright puts it like this: 'If the sponsor determines that it is essential for the project to be completed by a set date and there is no flexibility or slack in the project, then extra resource and extra cost might have to be accepted. It is the project manager's responsibility to find alternative methods and courses of action in an endeavour to keep the extra cost to a minimum.'[14]

Not all teams will respond well to this kind of openness. There is a risk that some departments might want to spend as little as possible to gain political 'points' and in doing so cut corners or report inaccurately giving you a false impression of what has been completed. On the other hand, some team members might want to appear important and report that their workstream has overspent due to fixing a difficult problem or working extra hours.

There is a further risk that being transparent with the budget situation could lead to money being taken away from your project – especially if you are managing your costs effectively. This is very dependent on the culture of your organisation. If it does happen, try not to take it personally. It is all to do with what the company thinks are its strategic priorities and unfortunately your project might not be one of them any longer. A strong Project Management Office (PMO) can help here, as well as a mature project management culture where people understand the value of not changing their minds about which projects to do every five minutes.

Salary and contractor costs are sensitive information. How can you share the entire budget with the team without divulging this sensitive information? You can't. Sharing capital expenditure – buying things – is different from sharing operating expenditure – how much it costs to run people and business as usual. To be honest, not many project managers have access to salary information for the people on their teams anyway. That's normally data belonging to the line manager. You may have day rates set by the company for different types of resources, which are not the same as salary figures. These will be publicly available, and you can use these in your forecasts and presentations to the team.

Projects cost what they cost, and if the investment is worth it and you haven't padded out the budget unnecessarily, you shouldn't have to worry about sharing the figures. However, it is up to you as the project manager to gauge the reaction of your team and decide how much openness is appropriate.

When you are reporting your budget to the team remember that in general people don't keep track of what they have done or how much time they have spent. How many times have you got to the till in the supermarket and not realised how much you have spent until the assistant asks for the cash? Your team may not realise how much of the available money they have already used up and sharing the budget with them through regular team meetings is a useful way of educating them about project finances.

While it might feel natural to keep the figures to yourself, sharing the project's financial status with the team on a regular basis can aid discussion and help facilitate collective responsibility for bringing the project in on budget.

7 AGREE WHO HOLDS SIGNING AUTHORITY

Signing authority is the ability to say 'yes' to expenditure. A project budget describes and tracks all the things that need to be purchased, but someone actually has to sign the invoice, raise a purchase order and be accountable for the money that leaves the company. It could be you as the project manager or someone else; either way when you buy something you need to know whose desk the paperwork should land on.

TAKING RISKS

Marie-Hélène Dupleix, a project manager from Brittany, was working on launching an online employee satisfaction survey for six thousand staff across five office buildings when she hit a problem. 'The budget was already approved, but I needed someone to raise a purchase order for a new server. The technical team said we needed it quickly to stick to the schedule, so I approached my sponsor. He said it wasn't his role.' Marie-Hélène spoke to all her key stakeholders and the finance department manager, but no one would take ownership of the budget. 'Asking around took me four days, and at the end of that we risked being behind schedule and I had achieved pretty much nothing except a headache,' she says. Marie-Hélène took a risk and found a template to produce what could pass as a purchase order. She added the project's budget code and sent it off to the suppliers. The server duly arrived.

'It took three weeks for the server to arrive, which was fine,' she says. 'During that time there were other day-to-day concerns and I really didn't give it a second thought.' But then the invoice came and the suppliers needed paying. Marie-Hélène approached her sponsor with the paperwork and asked him to sign it. 'He asked me if we were within budget and if this was planned expenditure. I said yes, so he signed it,' she says. 'I sent the forms to the finance department and as I didn't hear back from the supplier I assume they were paid.'

Marie-Hélène acknowledges this was a bit of a risk for her. The next time the project needed to spend money, she approached her sponsor again. 'He asked me what I had done last time, as he didn't remember authorising any expenditure,' she explains. 'I told him that I had raised the purchase order as we needed to move quickly. He said I could carry on doing that and signing for project expenditure as long as we were within budget, so I did.' Marie-Hélène feels she was lucky that it turned out this way, 'but sometimes you have to take risks to get results and this one really paid off,' she says. 'I got taken more seriously by my peers and the finance department, and I didn't have any more delays or headaches!'

The process for approving expenditure differs from company to company. A little research at the beginning of the project will let you establish the process you need to follow. There may be multiple steps in the chain, multiple signatures to get or committees to attend. Someone in the process may be very difficult to track down. The more process steps and the busier the people in the chain, the harder it will be to purchase anything. The benefit of finding out the process at the beginning of the project is that you can build adequate lead time into your plan. Knowing that it will take three weeks to get the three signatories in a room together with the right paperwork means you can make sure you start early enough to avoid holding up the rest of the work – or come up with more creative solutions like electronic signatures and online approval workflows.

'The ideal situation is for the project manager to have control over the budget,' writes Kim Heldman in *Project Management JumpStart*. 'This doesn't mean that you should have unlimited signing authority but you should be able to sign for normal supplies, contractor invoices, and so on.'[15] Heldman advises project managers to agree a certain level of signing authority for themselves with the sponsor and finance department. This avoids having to go through the purchasing process for small amounts of expenditure and will certainly lessen the day-to-day financial headaches.

Giving project managers any degree of signing authority will be a no-no in some companies. Heldman believes that 'project managers who have no control over the budget should not be held responsible for budget mishaps'.[16] However, your sponsor may not see it that way and it would be unprofessional to ignore the budget completely just because you don't have the appropriate authority to sign off invoices. You can still track expenditure, log the requests for purchases and flag any concerns to your project sponsor along with your recommendations for how they could manage the relationship with the budget holder.

Ask for the authority to make purchases against the project budget and if this is not granted (or only granted in certain circumstances) establish the process for gaining approval for expenditure outside of your authority.

8 ARRANGE FOR A PEER REVIEW

A peer review is an informal audit that looks at the project so far through a pair of external eyes. Peer reviews are a useful tool to check the project is on track and give you confidence that you are doing the best possible job. They can be run by an external company, the internal audit or quality function of your company or by the project management department with project managers carrying out peer reviews on each others' projects.

If you are offered the chance of a peer review, take it. If not, think about organising one for yourself at key points during the project: insights from an unbiased evaluation really will be worth it.

PEER REVIEWS IN PRACTICE

'Typically my participation in a programme review has come from either an unfavourable review by a US government agency review or an equally unfavourable internal management review,' says Bob Rodgers, Senior EVM Consultant at Virginia Beach-based firm Ten Six Consulting. 'As such they've welcomed our independent assessment which can reinforce the negative feedback they've received, but we also bring suggestions as to how to rectify problems or to develop recovery plans to improve programme performance.'

Bob carries out periodic reviews with Department of Defense contractors, after the government agency responsible for procurement has highlighted that they have not achieved the required standards during an audit.

'The process is to sit down with the programme manager to get their assessment of programme status then to confirm that their data submissions did in fact verify their assessment,' he says. 'If they don't, I circle back to understand the differences and once that's done, validate that internal reports.'

Validating the documentation can take some time. Bob typically asks for a suite of paperwork including the statement of work, work breakdown structure, risk log, master schedule and the critical path, responsibility assignment matrix, contract performance reports, float analysis results, and budget.

> Having studied the baseline documents and the programme paperwork for the last three months, Bob reviews the budget and the schedule to check that the critical path has no negative lags, negative float or open loops. He then selects the 'most errant' control account (Earned Value Management speak for chunk of work) to review further.
>
> 'I validate that the work authorisation document for the control account in question matches the performance reports and the responsibility matrix, and then I audit that control account to confirm that budget, scope and schedule all check out,' he says. This review also covers any potential critical path activities that impact the work. 'Then I review the findings with the relevant manager and recommend corrective actions.'
>
> Bob's peer review may include several areas of the project and when it's complete he assembles the findings and reports back to the programme manager and his or her team. 'It's hard to estimate the time it takes with the programme team as it's a function of how many problems exist,' he says. 'If there are only a few or none a review can be done in three to five days, otherwise it can take two or three weeks. My longest interview with a manager to date was nine hours. Needless to say it did not go well, but we found a number of items that helped the individual concerned to improve performance over the long haul.'
>
> Bob has some advice for project managers carrying out peer reviews on other projects. 'Always dig into those control accounts that have habitually incurring cost variances and schedule variances,' he says. 'There's a reason and they don't need to wait till the hammer comes down on them from on high to be aware of it and take remedial actions. Be proactive in resolving recurrent problems.'

A peer review is:

- a process;
- an unbiased, friendly assessment of the project activity;
- going to point out areas where you could improve the management of the project, and;
- a way to get suggestions on how to make things run more smoothly.

A peer review isn't:

- a quick fix;
- meant to identify mistakes;
- going to parcel out blame, or;
- intended to reduce the morale of your team.

There is no single best time for a peer review to take place. Some project managers recommend early in the project to check that the initiation phase has given the project a good foundation. Others would suggest just before the project is implemented. A review towards the end of the project will give you a clear summary of progress to date and actions to put in place to ensure the project completes successfully.

If your PMO has no clear guidelines about the timing of a peer review, schedule one at the time that makes the most sense to you and your project. Too early and the reviewer will have nothing to look at. Too late and you won't have the opportunity to act on the recommendations.

Continuous improvement is part of the project manager's role so where is the value in a peer review? Projects don't get into trouble overnight, and often the project manager can be so close to the detail that it becomes hard to see the issues objectively.

> During a peer review expect to hand over copies of your monthly reports, your budget spreadsheet, plans and other project documentation. Have all this to hand before the review starts. Free up some time so you can talk to the reviewer and make your team available too if necessary – a thorough review can take a while and you need to co-operate with the reviewer to get the best results.

A peer review will normally look at the project holistically, but if you have specific areas of concern then ask the reviewer to pay particular attention to those. This is especially useful for budgets where a second opinion will either reinforce your methods or identify ways in which you can improve the tracking of project financials. If you don't have much experience of Earned Value Management, for example, choosing a reviewer who can give you some constructive pointers or reassure you that you are handling the models correctly can be a big help. A review of your project budget can also give your sponsor confidence that the money is being handled appropriately while at the same time pointing out future shortfalls.

The reviewer should produce a final report including their forecasts for the total expenditure and predicted finish date, based on their assessment of the project. These forecasts will either support your own or give you an alternative to consider. Any alternative view of your project forecasts should be backed up with some well-thought-out reasons which will make it easier for you to decide whether or not you want to make changes to your plan based on the reviewer's report.

> **HELP! I'VE BEEN ASKED TO BE A PEER REVIEWER!**
>
> That's great! Be flattered if someone asks you to review their project. Someone thinks you are a good enough project manager to pass judgement about another project, objective enough to present your findings in a clear and blame-free way and detailed enough to uncover things that they themselves might have missed.
>
> Being a reviewer is not a daunting task. Find out if there are standard templates for carrying out peer reviews already in use by other project managers. Even if your search doesn't turn up anything official, speak to someone who has done

it before. Think about what you would want to know if it was your project: is the budget on track, is the schedule realistic, how are risks and issues managed? You do not need to be an expert in the technical subject matter of the project to carry out a successful peer review but there is an obligation to give your feedback in a constructive way. Focus on constructive criticism and suggestions for improvement rather than highlighting errors the team can now do nothing about.

A peer reviewer will produce a final report of their assessment. This is a great start, but alone a report will not change how you manage the project. To make the whole review exercise worthwhile you really need to put those recommendations into practice. However, you may be surprised at the recommendations: you won't always be faced with a long list of suggested changes. Todd Williams, in his book *Rescue the Problem Project*, puts it like this:

Doing nothing based on ignorance will most likely result in failure. However, performing an audit or analysis of a project and then making a decision to make no changes is an action – a conscious decision to let the problems resolve themselves. This is a valid choice. The best example of this is when a project is incorrectly declared to be in trouble; for example, a set of people (say, other managers or a group of politicians) may claim a project is in trouble for their own reasons.[17]

More often than not, though, your impartial reviewer will have some suggestions about how you can manage the project more effectively. Carefully read through the recommendations and work out how and when they can be implemented: obviously the sooner the better. Ask the reviewer's advice if you cannot see how to turn the recommendations into reality. They will have suggestions for a successful implementation which might help you see the problems in a different light.

Schedule peer reviews at sensible intervals during your project and be sure to act on the reviewers' recommendations.

9 MANAGE PROJECTS WITH NO BUDGET CAREFULLY

While most of the projects that we read about in the press have huge budgets, there are a lot of projects at the other end of the scale. The majority of small projects, especially those without IT involvement, are run without a specific budget, around the edges of a manager's day job. Any costs have to come out of the business-as-usual provision which means there is no allocated project budget. They may not have a specific project manager managing them. Another scenario is when a project manager is assigned to the work, but the budget is still held by the business department and is not given to the project manager to handle – most likely because there is no set amount to spend and again it has to come from business-as-usual spending. So when your project has no budget, how can you keep a budget on track?

NOT COUNTING THE PENNIES

'When our branch manager told me I had to set up an online system to run the quarterly quality assessments for call centre staff with no budget, I admit I wasn't happy,' confesses Gordon Harvey, a senior IT project manager at a travel firm. 'Partly because I didn't think it would be possible and partly because I'd just finished running a project with a budget of £100,000 so I felt it was a bit beneath me.' The company's customer-facing staff undergo a rigorous assessment every three months to check they are providing a top-quality level of service to travel customers. This process took up a lot of time for team leaders and the paperwork involved made everyone dread test week.

Gordon thought creatively about what could be an appropriate solution and drafted in a part-time team member from IT to help. 'We were a project team of two: me and one other, and we were both only working on this project two days a week,' Gordon explains. They needed to find something simple, quick, and more importantly, free. The obvious solution was to take advantage of the company intranet, but it had never been used in that kind of way before. Gordon was unconvinced that his IT colleague had the skills to build something complementary. 'By chance I was talking to someone else in a different branch about a different project, and he happened to mention they were in the process of building a database to hold customer satisfaction scores,' Gordon says. He immediately saw a link between the customer survey data and the type of service questionnaire the call centre agents completed four times a year. The database was being created in conjunction with the marketing team who were managing communications to the customers. Gordon put a call in to the internal equivalent – the Internal Communications manager. 'She was interested and helpful in a reserved way,' he says, 'but didn't have any resources available to help practically. What we really needed was a comms professional who could spend some time with the marketing people from the other branch to find out how we could apply the same logic to the customer service questionnaire.' Gordon got on the phone to his sponsor and explained the difficulties they were having. Some negotiations took place at a senior level and Gordon found himself with the comms person he wanted working on the project for a day a week. In return, he later found out, his sponsor had promised to move the resolution of an outstanding technical issue that was affecting the Internal Communications team higher up the priority list. Gordon got the specialised resource he needed and Internal Communications benefited too: no need for money to change hands or resources to be formally 'purchased' and the project stayed within the 'day job' confines originally prescribed – only with a little extra help.

Projects with no specific financial amount attached to them are normally expected to be delivered using just the resources available as part of a day job. That basically means drawing on the people around you to do whatever it is that needs to be done. Having a project with no budget takes away some of the financial headaches but doesn't mean you are in for an easy ride. You will still have deadlines to meet and requirements to deliver.

In some respects, no-budget projects are harder to deliver as you cannot throw money at a problem to make it go away nor will your team have access to overtime payments if things start to slip.

If your sponsor hands you a project and then says, 'there's no money available to do this', count to ten and try to avoid spitting out, 'You must be joking!' Ask them how they expect it to be achieved. A sponsor who is serious about a project will already have thought about what they consider to be a reasonable investment for a successful delivery. Take them through the questions below and start to work out where your boundaries are, particularly in relation to the people you have access to and the amount of time you and they can be expected to spend working on the project.

SPONSOR QUESTIONS FOR NO-BUDGET PROJECTS

- Am I full-time on this project?
- If not, what percentage of my time do you expect me to spend on this?
- Do I have any full-time resources?
- If not, what percentage of their time do you expect them to spend on this project?
- For any resources not under your control, has their manager agreed that they will be working on this project?

- Can any costs come out of the business-as-usual budget?
- To what limit?
- Who will authorise this?
- If the business-as-usual budget is not available, how do you want me to deal with unforeseen actual expenditure?
- At what point does the project become unfeasible?
- When does the resource investment become too much for your intended deliverables?

Once you have a clear idea of where your sponsor believes your boundaries are in terms of consumable resources, both business-as-usual budget expenditure and time, you can begin to work on the project within those constraints.

Make sure any assumptions or constraints are written in your project initiation document. For example:

- the business-as-usual sales budget will cover the cost of reprinting a new edition of our catalogue;
- the schedule has been produced assuming that no overtime is available;
- all resources will be available as necessary;
- the system changes can be achieved using the maintenance budget.

When you are not 'buying' your resources formally, and they work for someone else, there is a risk that their own day job will take priority over the project. Despite good intentions, there will be times when staff shortages, increased workloads or other short-term crises drive your team back to their normal activity. If you can, schedule contingency time to keep your resource planning flexible. Always include an item in your risk log about the possibility of resources being pulled off the project. As a minimum, each time you review the log it will prompt you to look at the current situation and see if you need to take any action.

If at any time the project looks like it will have to spend real tangible money and you don't know where it will come from, raise this immediately with your sponsor as an urgent issue.

Even if no specific budget is available for the project, clarify with your sponsor what they consider to be a reasonable 'investment' in terms of time and work within those documented constraints.

10 USE TIMESHEETS FOR TRACKING TIME

Project recording normally takes the form of timesheets, completed by the project team on a weekly basis. Time recording is a seemingly straightforward and useful task but it can alienate team members if the rationale behind it is not explained and the methods are applied too rigidly.

INTRODUCING TIMESHEETS

Timesheets were introduced at Gemma Viles' company as there was no visibility of what people were working on. 'Originally the project objectives were to install an enterprise project management system to allow better visibility of the portfolio, so that it could be managed and resourced, tracked and controlled,' she explains. 'The project was "hijacked" by the director as there was a push from senior management to account for people's hours. Although timesheeting was included in the original scope it was intended on being a much later deliverable than it actually was!'

As a result of the director's involvement, the focus of the project changed with the new objective being to account for the hours of all IT staff to enable reallocation to customers. Timesheets were a critical part of that. 'It has been a bit of a struggle,' Gemma says. 'Encouraging people to book at all is hard enough but when you need them to book accurately it's even harder. We've had some issues such as people consistently booking leave as statutory holiday when there have been no Bank Holidays in the week. Also when we started the amount of Admin time booked was disproportionately high.'

Helping line managers understand the importance of booking and of booking accurately has helped get better data from the timesheets. 'Once they understood that timebooking helps us justify (or even increase) our headcount they were on board,' Gemma says. 'It's been harder with individuals as filling in a timesheet is much less important on a Friday afternoon than going home! Again, an understanding of why we need their help and how this impacts the customer has really helped. As has the persistent "poking" from the PMO, going as far as sitting with people and showing them how to complete a timesheet, which is a very simple task.'

Gemma's department has found benefits from insisting on timesheets for project teams. 'We can now reallocate costs back to our customers,' she explains. 'This means that where our IT budget was stretched and we were looking at a huge overspend at the end of the year, it is now much healthier with major chunks of

> project work being captured and charged for. We can account for time spent on supporting specialist applications which could help us to retain posts going forward.'
>
> Admin and absence time is now only around 15 per cent of logged time, the rest being chargeable. 'Now that timesheeting is the norm, our project managers are more aware of their project budgets and there is more communication with their customers around spend, resulting in a greater financial awareness,' Gemma explains. 'Going forward we are encouraging timebooking at task level and we ultimately aim to have plans driven by the submitted (and accepted) hours from timesheets.'
>
> For all the benefits, the change in the direction of the original project meant there were some costs. 'Some of the project related objectives were lost, and we are rectifying these now,' Gemma says. 'In hindsight the implementation of timesheets over project functionality was a bad move. It has given us what we wanted but there has been some rework to pull things back to a project focus. The benefits of timesheeting though have probably been worth the sacrifice.'

Timesheets allow you to monitor the work of your team. In its own right, this is a vaguely useful activity. In a world where as a project manager you need to be able to justify what your team has been doing, and maybe cross-charge another department or organisation for their time, timesheets serve as a method on which to base invoices. They are certainly not a perfect way of monitoring activity, but as no one has yet come up with an accurate and foolproof alternative, many organisations rely on them. Time recording can be used for more than just checking up on your team or satisfying internal accountants.

You can compare the timesheet data with the original estimates in the early stages of the project, and see how closely they match. A discrepancy will show you that the estimates are inaccurate. 'For example, if you have used 30 per cent of the allotted time and you are only 10 per cent complete, that is a red flag,' writes Curt Finch in his book *All Your Money Won't Another Minute Buy*.[18] 'Isn't it better to have that red flag raised when you have spent 30 per cent of your money rather than 80 per cent? That's the difference that continual project tracking makes.' Doing this comparison exercise early on in the project can give you an insight into any potential overrun you may encounter in the future.

However, getting accurate data from your team can be difficult. Their timesheets may routinely show that they spent eight hours a day working on your project, but that implies they took no toilet breaks, answered no phone calls and never responded to any non-project emails, which is highly unlikely.

The problem is exacerbated if the team doesn't understand why you are collecting the data. Timesheets can actually put a barrier between the project manager and their understanding of what is happening if team members think they are being treated as part of a number-crunching exercise and not as intelligent, valued members of the project team who can effectively manage their own workloads.

> A project manager spends 10 per cent of their time chasing progress reports, which is about 16 hours a month. Team members spend 12 hours a month reporting their progress. For a medium-sized company, this equates to about £600,000 wasted a year on tracking down information.[19] Using online tools for real-time task tracking and effective reporting mechanisms can turn some of these hours into productive time.

Explain to the project team how you are using the data to improve estimating, cross-charge other departments and justify the amount of people working on the project (or anything else that you use it for). This should encourage them to be realistic in their reporting and more comfortable sharing the information regularly. Frequent discussions about the project schedule will help your team feel confident that you understand what they're doing. If they feel strongly that you appreciate and are responsive to their difficulties at a task level this will create less of a reason for them to submit inaccurate timesheets, especially as they know you are aware of any discrepancies.

> Time recording can be a useful activity, but don't let it be the only way you monitor your team's activity or you risk being fed inaccurate data week after week.

11 BUDGET FOR CHANGE MANAGEMENT

For every £1 spent on change management on large projects, organisations get a return of £6.50, according to research by Changefirst,[20] a company that helps organisations implement change more effectively. Of course, you would expect a change management company to advocate change management, but their study shows massive returns on doing change management properly. Survey respondents were asked to look at projects six to nine months after completion to see whether the change had 'stuck'. The results show that on large projects, over 40 per cent of change managers believe that at least a third of the financial gains could be attributable to change management activities alone.

Not all projects have dedicated change managers – in fact, you are more likely to find change managers on business change programmes than on projects. As a result, project managers have to take on the change management activities as well. The Changefirst research still stacks up for small projects: on these, every £1 spent on change management generates a return of £2, which is still impressive, especially when you think that dedicated change management budgets are just a small part of what it costs to deliver a project – typically change management activities cost around 10–15 per cent of the overall budget.

CUTTING OUT CHANGE MANAGEMENT

'I was working on a large programme to implement ERP software in a UK firm,' says Heather, a programme manager from London. 'The budget spanned two years with implementation of the software in three pilot sites during the first year, and the other offices and business units due to get the software in Year Two.'

The budget for the first year included costs for change management activities. 'We set up a wiki for frequently asked questions about training, built a new intranet site and produced a programme newsletter,' Heather says. 'We also scheduled monthly briefing and change management sessions for key business people, held regular conference calls and twice during that first year brought together a cross-business team of local experts who would be responsible for embedding the change back in their departments.' Not all those activities cost money – the newsletter, for example, was distributed by email, so (apart from the time it took to put together) was effectively free. The most expensive change management activity was training.

'We held webinars and face-to-face training sessions,' Heather explains. 'The training sessions covered both technical training and process training focusing on the "how do I" elements. We also ran "train the trainer" sessions and workshops for those people who would be leading others through the changes. Together, this was our most significant investment in change management that year as we needed to convert four meeting rooms into classrooms, complete with PCs, printers and other specialist equipment.'

The results from the change management workshops and training sessions were positive. Feedback from the attendees was good and in the main people left the sessions feeling confident about the forthcoming changes and how their jobs would be different once the new software was implemented. When the software was launched at the three pilot sites there were some initial problems to sort out but generally the teams coped well. 'We were able to test out our change management approach as well as the software itself,' Heather says. 'Our pilots were really valuable in that respect.' Heather and her team were able to tweak the change management approach and the workshop contents so that the other business units would benefit from what they had learnt during the pilots.

Unfortunately, in the second year of the programme the budget came under tighter scrutiny. 'As a result of some technical problems, we needed to make savings,' Heather explains, 'and change management was one of the things that was cut.' The team lost their budget for classroom training. 'As a result, we are trying to achieve the same amount of change management through webinars and online learning,' Heather says, 'as we can't pay for trainers or delegates to attend face-to-face sessions any more. It's a huge disappointment as we know that they worked.' Heather doesn't believe that enough time has passed yet to be able to judge the results from the software implementation to other business units but feels that the overall outcome will not be as good. 'Change sticks when you invest in it and invest in the people who are expected to change,' she says. 'If you can't do that, you can't guarantee the results, and that makes whole projects a waste of time.'

To understand how much change management activities will cost, you need to first work out what they are. If you don't have a dedicated change manager on your project or programme, this responsibility will fall to you. Work with the team to brainstorm ideas for what change management tasks will be required on the project, and then you can allocate costs to them.

Change management doesn't have to be expensive, as this case study and the Changefirst research shows, but you should make some financial provision for it. Even if you aren't sure what change management tasks will be required at the outset of the project, you can still put in a line on the budget of around 10 per cent of the overall budget to account for anything that you subsequently find out that you do need to do.

Don't forget to budget for change management activities, as these will pay off in the long run by ensuring that the change delivered as part of your project has a higher chance of 'sticking'.

12 UNDERSTAND THE BENEFITS

We often think of project benefits as financial. Benefits are couched in the language of return on investment, payback period and increased revenue. But benefits don't have to be simply about improving the company's bottom line.

FINDING THE BENEFITS

Paul was tasked with replacing the outdated X-ray equipment in a private hospital with new digital imaging technology that would allow radiologists to see X-rays on computer screens instead of on photographic film. 'The project involved removing all the film processors in the hospital and upgrading the X-ray equipment to accept digital plates,' he explains. 'It meant a number of changes across the hospital, not least for the medical records team who used to store thousands of X-rays on film. The aim was to have all the new records stored digitally.'

The project cost hundreds of thousands of pounds for very little tangible financial return. 'It was pretty difficult to work out if we actually made any money from upgrading the kit,' Paul says. 'The benefits were far more about patient care and consultant satisfaction, plus staying competitive.'

Medical technology advances quickly, so it was important for the hospital to be seen to have some of the latest equipment. As other hospitals in the region were also upgrading, it meant that it would be easier to transfer X-ray images between locations as well, which was useful for consultants who practised in a number of different hospitals or who needed to share patient records with colleagues to ensure continuity of care. 'It is much faster to process an X-ray digitally than to develop an image on film,' Paul explains, 'so patients get their results more quickly, and sometimes they are available even before they have walked back to the consulting room after having the X-ray. But if we had evaluated the project purely on financial grounds we would never have done it.'

There are seven areas where you can look for benefits on your project.

1. **Mandatory benefits** Mandatory benefits could be a legal or policy requirement related to something that is mandatory for your organisation. This type of benefit enables you to satisfy legal, compliance or other mandatory requirements that are essential to keep the company operating. Mandatory benefits often make

the business case for your project stand by itself – you won't need to look for any other types of benefits. However, it is worth evaluating the project to see if there are other benefits to be had as this will show that your project will provide some added value.

2. **Quality of service benefits** Quality of service benefits result in providing a better service to customers. This could be through a shorter turnaround time for queries or reducing customer complaints. Ideally, these benefits should be something quantifiable.

3. **Internal benefits** This type of benefit helps improve things like corporate decision making. These are internal improvements that won't be felt outside the organisation but can streamline or support management processes.

4. **Flexibility benefits** Flexibility might not be the first type of benefit that comes to mind when assessing how a project will help a company. It is difficult to measure, but there are benefits to be had from many projects around ensuring the organisation continues to respond well to change without incurring extra costs. Some examples of projects that have flexibility benefits include an initiative to introduce home working or to implement new change management processes or software.

5. **Risk management benefits** Risk management should be undertaken on every project, but it can also be one of the project's benefits. Risk management benefits are those that help a company reduce its risk by being better prepared for the future. Depending on the project, these could also support a business case alone, even if the costs of doing the work are high. For example, redesigning your disaster recovery strategy might be an essential piece of work to mitigate business risk, even though you won't get a financial return on the project.

6. **Productivity benefits** Projects nearly always have an impact on people, and projects that improve staff morale or better motivate teams can possibly claim productivity benefits. Other types of productivity benefits could be measures to reduce staff turnover, introduce reward and recognition schemes, or collaboration software. Think carefully about how you will measure productivity benefits as they can often be quite difficult to quantify if you do not have good baseline data.

7. **Financial benefits** Don't forget the benefits that are driven by cost. Financial benefits can be gained through reducing costs while maintaining quality, or through delivering increased revenue, for example through launching new products or reducing recruitment costs. You can also include financial benefits from keeping revenue the same but getting it faster, as these will have a financial impact on your profit and loss as well as on timescales.

It's very important to understand how your project is contributing to the company's objectives overall by delivering its benefits. Starting out with a clear idea of what benefits will be achieved by the project makes it possible for you to gather the appropriate data, report accurately and put in place systems for benefits realisation to go ahead in a constructive way.

On many projects the project manager does not get involved until the business case is prepared and approved. As a result, it is likely that you will 'inherit' the benefits profile of the project instead of being involved in defining it. That's not a problem – you can still read through the business case to familiarise yourself with what the original team thought the benefits would be. When you know what they are, you can work with your project team to deliver them.

 Even if you are not directly involved in producing the project's business case, take time to understand how your project delivers benefit to the company.

FURTHER READING FOR THIS SECTION

Bradley, G. (2010) *Benefit Realisation Management: A Practical Guide to Achieving Benefits Through Change*, 2nd edition. Gower, Farnham.

Callahan, K. R., Stetz, G. S. and Brooks, L. M. (2007) *Project Management Accounting*. Wiley, Hoboken.

Elkjaer, M. (2000) 'Stochastic budget simulation'. *International Journal of Project Management*, 18, 139–147.

Fleming, Q. W. and Koppelman, J. M. (2005) *Earned Value Project Management*. PMI, Newtown Square.

Frow N., Marginson, D. and Ogden, S. (2005) 'Encouraging strategic behaviour while maintaining management control: Multi-functional project teams, budgets, and the negotiation of shared accountabilities in contemporary enterprises'. *Management Accounting Research*, 16, 269–292.

Portny, Stanley E. (2001) *Project Management for Dummies*. Wiley, New York. Appendix B (pp. 319–329) has an excellent introduction to Earned Value Analysis, including easy-to-follow worked examples.

Rasmussen, N. and Eichorn, C. J. (2000) *Budgeting: Technology, Trends, Software Selection and Implementation*. Wiley, New York.

Webb, A. (2003) *Using Earned Value, a Project Manager's Guide*. Gower, Aldershot.

SECTION 2:
MANAGING PROJECT SCOPE

INTRODUCTION

A man steers well who reaches the port for which he started.

Lucius Annaeus Seneca (c. 4 BC–65 AD), *On Benefits*

Scope forms part of the 'golden triangle' of project management (also known as the triple constraint) along with resources and time. Figure II explains the triangle. Scope, resources and time are equally balanced within a project and form an equilateral triangle. If there is a change to any part of the triangle, another element has to change as well to keep the balance. For example, if scope increases, time has to increase to give the project team longer to deliver the proposed changes. If this is not possible, more resources have to be found to do an increased amount of work in the same amount of time. As with all theoretical models, the practical application is perhaps more pragmatic. The 'golden triangle' assumes quality is a constant but a pragmatic solution to achieving things faster and more cheaply is to deliver to lower standards of quality. You'll see various versions of this triangle including time/cost/quality as the three points and quality/risk/customer satisfaction, and sometimes a combination of these factors amalgamated into a hexagon.

Figure II The golden triangle of scope, resources and time

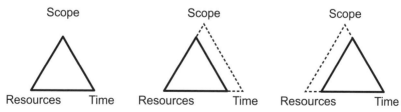

In the classic version of the triple constraint as illustrated here, scope is perhaps the most fluid of the three as changes are an inevitable part of project management. The average project goes through four formal versions of scope and ends up only achieving 93 per cent of what it set out to deliver.[21]

This section looks at why scope management is so important and what project managers can do to avoid the traps of assumptions, implied organisational knowledge and scope creep.

13 KEEP IT SMALL

It is very tempting at the beginning of a project to include loads of things in scope and plan a 'big bang' deployment. However, this is rarely the best way to deliver change, especially in the IT arena. Where possible, packaging the change up into smaller, interim deliverables supported by full testing is a safer and more robust approach to implementation.

> **CREATING A SOCIAL WEATHER COMMUNITY FROM SCRATCH**
>
> Paul Craig took the role of Scrum Master on a project for the UK Meteorological Office. The Met Office provides weather data to NASA, the US Army and other places that cannot get the information from their own national met service. 'They are one of the global leaders in terms of a national meteorological service,' he says, 'but they were suffering from a bit of a PR challenge.' For all their skills, data and links to people in high places, they are still known as the organisation that gets the weather wrong. Everyone remembers the time when Michael Fish didn't predict the hurricane of 1987 (and if you don't, YouTube has the moment memorialised).[22]
>
> 'There are thousands of amateur weather enthusiasts around the globe and every week someone would knock on the door in Exeter with 20 years of collected data and offer it,' Paul says. Until recently the Met Office couldn't do anything with it. Then a new IT Director arrived and wanted to start harnessing community weather recordings. He set the team a challenge: to see if new technology could be used to increase and improve the collection of weather data outside the Met Office. An Agile project team was put together and tasked with delivering something useful in the next 12 weeks.
>
> 'It was the first major Agile delivery project for the Met Office and this was completely new for them,' Paul explains. 'We had to force ourselves to slow down for the first 3 weeks. Within a month everyone got it.' Knowing that the team had to start small, they wrote down all the potential requirements and allocated each requirement 'requirements points'. 'The triple constraint still applies in Agile,' Paul says. The project had a fixed end date and a fixed cost so it was the scope that had to change. The project team held a two-day workshop to go through all the requirements. 'We said, "With your budget you can afford 300 requirements points. What do you want to do?"' he explains. This gave the team a straightforward way to put together a small but comprehensive scope, and meant they could easily justify why some things were in and some things were out.

That was just as well. The plan was to start small but to scrap the project completely if it didn't deliver anything that people wanted to use in three months. In the summer of 2011, WOW (Weather Observations Website)[23] was launched. 'All the blockers you'd normally find on an IT project have been removed and it's very cost effective,' Paul explains. The cloud-based architecture using the Google App engine allows the development team to get on with what they are good at without having to worry about patch levels, servicing the environment and so on.

Despite the worry that short timescales and a limited project scope would mean that this small project wouldn't deliver anything of value, Paul says that the amateur weather community jumped on board almost immediately. 'The community is self-policing,' he said. 'They assess and review each others' data so they guarantee good quality.'

Using open source platforms also meant that the community could get involved with developing it. Someone wrote an add-on which enabled the site to display moving pictures, which was an improvement the project team hadn't thought of (or hadn't considered for the first iteration). 'It was a fledgling idea and it snowballed,' Paul says.

The WOW project created a brand new 'social weather' project from nothing. Since the launch, the site has received visits from 150 countries and receives one million weather reports a week from 3,000 to 4,000 amateur weather observers around the world – and they have done this all with no marketing.

Did it meet its objectives? WOW was never supposed to make money or to gather data to replace what the Met Office collects via the 'official' weather stations. The aim was good PR and Paul believes that this has been achieved. The amateur weather community likes it, the site has appeared on TV, and Google has produced a case study about it. For something that was a three-month experiment, WOW has taken the amateur weather community by storm.

The key to getting it right first time is keeping it small to begin with. Agile methods, with multiple iterations (or sprints) each adding more features to the product, lend themselves naturally to incremental build. You can also achieve this with waterfall methods if you keep to the principle of starting small.

When you plan the scope of your project, don't be tempted to include too many things. You can always have a phased approach where your first project delivery (and so your first scope statement) covers a small portion of what you would ideally like to do, then add subsequent phases to manage the rest of the roll-out.

There are very few projects that would not benefit from a pilot or proof of concept stage at the outset, so consider including one in your project scope. Rolling out a new system to a handful of users, asking them to test it rigorously and then making changes based on their feedback is a lot easier than deploying the system to everyone then managing the internal communications, bad publicity and general ill-feeling when you implement the fixes later. Your users will be asking themselves: 'Why would this latest update be any better?'

Piloting is also a way to contain costs. Identifying problems early gives you the chance to manage them when they are still small and can be corrected without significant cost.

While being able to test functionality and get early user feedback are normally the main reasons for a phased or incremental deployment, keeping it small also allows you to learn from the implementation process. The system might be great, but could you tweak the training material? That branch opening was a success, but when the bigger branch opens could you get more local press coverage? The new product is selling well to a trial group, but how can you act on customer feedback regarding service? Take the lessons from customer feedback about the end product and the feelings of your users with regard to the deployment process to inform and improve the next phase of the roll-out. Complaints are a gift: someone has given you the chance to tweak the product and get it right before you unleash it on your customers and to confirm that all the snags in the implementation plan are ironed out before you go for a huge launch. The benefit of starting small is that you have the opportunity to make things better for next time. Use it.

THE SIX STEPS OF PILOTING

1. Establish how a pilot could add value to the project by mitigating risk.
2. Get commitment from relevant stakeholders.
3. Define the evaluation criteria for the pilot: how will success be measured?
4. Produce a project plan and other project management deliverables.
5. Deliver the pilot: run it for a predetermined length of time.
6. Evaluate the success based upon the criteria established in step 3.[24]

Putting a pilot or proof of concept stage in the scope of your project and/or planning an incremental approach to delivering functionality will allow you to fully test both the deliverables and the implementation approach, so that you do not lose the support of your project customers by delivering something that is not fit for purpose.

14 KNOW WHERE YOU FIT

Your work on a project fits into a larger strategic picture for the organisation as a whole, or at least it should. Everything you do on your project should be aimed at helping the company achieve its strategic objectives, and it should be clear to you how you are contributing to this strategy.

LINKING EVERYTHING TOGETHER

Antony Tinker, Sales and Marketing Director for Kent-based change consultancy Innovative Edge, knows a bit about how to ensure that projects are linked to programmes and upwards to business strategy. After all, it's what his company does.

'As we are all about aligning people and strategy a big part of our exploration phase when we work with new clients is to understand their business strategy,' he says. 'If their strategy is clear and understood we will link everything we do to it, or parts of it. If it is ambiguous, not understood or not believed then we will propose a solution to address this.'

The exploration phase is where Antony and his team sit with the client and establish what project benefits they want to see. They spend a lot of time carrying out one-to-one interviews, exploring the commercial background and the company culture, asking about the rational and emotional issues and trying to get to the bottom of the benefits the company wants to realise. 'By identifying their pain points and having a discussion they can begin to articulate how life would be different (better!) without those pain points,' he explains.

Once the benefits have been established, they can be delivered through a project, or series of projects. However, each project still needs to link to a programme of work and the overall corporate strategy. 'When we present our proposal back to the client we cover off the features and the benefits of carrying out our proposed project plan. As we design every intervention bespoke for each client this is relatively easy to convey, and very authentic,' he says. 'Authenticity is vital. If you say it's about them then make sure it is. They'll see through fake words and actions very quickly.'

He is also prepared for things to change, as they often do on projects. 'You should be open minded – as should the client – to evolve your plans as you work through the project, as long as you remain dedicated to genuinely helping them achieve their objectives and goals,' he says. 'Inherently therefore everything is linked, no matter what changes, and we can demonstrate this.'

Antony believes that linking benefits, projects, programmes and corporate strategy results in better alignment between individuals and the company goals, happier teams and a positive impact on financial results.

'If you don't know how your project links to business strategy, ask more in breadth and depth questions,' he recommends. 'Make it your client's agenda and not your own. When you ask a question listen to the answer rather than thinking of your next question.'

The OGC P3O standard defines a portfolio like this: 'the totality of an organisation's investment (or segment thereof) in the changes required to achieve its strategic objectives'.

The OGC P3O standard defines a programme like this: 'a temporary, flexible organisation created to coordinate, direct and oversee the implementation of a set of related projects and activities in order to deliver outcomes and benefits related to the organisation's strategic benefits'.[25]

Have you ever asked yourself why you are working on something? If not, then now is the time to start asking. Often project managers are told what projects to work on with very little background or context. As a result, it can be hard to energise a team towards a common objective when you can't see how your work contributes to the bigger picture.

Ideally, business strategy should be manifested in the form of portfolios, which are delivered via programmes and projects. Therefore it should be a straightforward exercise to map your project through a programme and portfolio and back to strategy, as you can see in Figure 14.1.

Figure 14.1 The strategy triangle

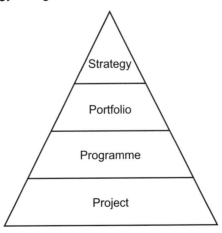

Durbin and Doerscher, in their book *Taming Change with Portfolio Management*, describe different types of portfolios and explain the links between project portfolios and programmes/projects like this:

> Execution portfolios are used to manage the tactical details of how projects, services, and the activities within them are planned and executed as well as how resources are assigned and managed. Execution portfolios give managers the opportunity to organise and view data from common information sources based on their specific decision rights and needs. Managers can focus on the information and issues that are meaningful to them and coordinate their activities with other managers. Because of the relationships defined by the information structures, they can also clearly see how their actions relate to the bigger picture.[26]

However, many smaller companies (and even some large ones) don't work on a portfolio basis and use programmes and projects directly to deliver strategic objectives. Even if your company doesn't have programmes of work, you should still be able to tie your project back to something bigger than the output that you have been asked to deliver. If you are struggling, see if your project's objectives align with your sponsor's personal objectives for the year.

If you can't see any clear link between what your project is supposed to do and how it will help the organisation move forward, maybe it's time to close the project down. An example of this would be where the business strategy has changed since the project's inception and the project will no longer deliver a valuable output. This could happen if there is a change of senior management or a corporate refocusing initiative. Suggest to your project sponsor that if your project is no longer going to achieve anything that supports the overall business strategy that it should be stopped. The resources could most likely to be put to better use working on something that will move the organisation forward.

> You and your team will feel as if your work is more worthwhile if you understand how it contributes to the overall aims of the company, and it makes better business sense to ensure that your project is aligned to strategy – so find out!

15 WORK OUT HOW TO MANAGE CHANGES

Change is an inevitable part of projects. The purpose of a project is to change something – otherwise we wouldn't do projects. There are two types of changes:

- changes that come about as a result of the project i.e. the deliverables of your project, and;
- changes within the project e.g. as a result of factors like new requirements, upgraded technology or external pressures such as new laws or modifications to regulatory frameworks, or a sponsor who changes the priorities.

The impact of the first type of change is hopefully the subject of your project and has been carefully planned with a communications and implementation strategy. It is the second type of change with which we are concerned here: how to manage changes that affect your project objective in some way. Changes will happen, so working out in advance how best to handle these will make the whole process easier when one comes along.

> **KEEPING IT UNDER CONTROL**
>
> Process-driven industries rely on accurate monitoring and careful management of changes, as Leo Dijkhuizen, knows well. He has spent ten years in a chemical and pharmaceutical environment in the Netherlands, where strict change control procedures are a necessary part of his role as a project manager specialising in SAP software implementations. Working on projects with budgets that stretch into millions of dollars and time frames of several years he has found it absolutely critical to manage changes in a controlled way.
>
> Leo explains the change control process of a project to define and implement a global SAP system: 'Each change had to be justified with a business case,' he says. 'The steering committee reviewed and approved – or rejected – each business change.' The system was being rolled out across six European countries and the US, with the aim of globalising the company's supply chain and lowering inventory costs.
>
> The project took three and a half years, and with a budget of $32 million it would have been easy for small changes to mount up over time and push the scope and budget way over what had been anticipated. 'The customer required a tight control

> over changes, so it was not difficult to implement a change control process,' Leo says. Having a global, centralised change control process helped to achieve a smooth roll-out across the seven countries. The steering committee signed off each change and was sure they individually each understood the implications.
>
> Leo's advice is to find someone in the project structure who can act as a sponsor for the change process. 'Make sure this person operates at the right level in the organisation. If you cannot find someone who meets that description, ask the steering group what their policy is regarding changes and develop a process to manage each change around that.'

Change control or change management is the process of managing unplanned but desired influences on the project. It is important because any change will:

- need to be analysed for its impact on project objectives;
- need to be analysed for its impact on project scope;
- modify your existing plans, and;
- need to be recorded properly for a complete audit trail.

THE FIVE STEPS OF A CHANGE MANAGEMENT PROCESS:

1. A request to make a change to the project is received.

    Requests are handed to project managers in a number of different formats, regardless of what the official process is. Your company may have a specific change request form or online template, but you might receive the request via email or during a conversation. Ideally, you want the person suggesting the change to be as specific as possible and to put their request in writing, preferably using the online change workflow, if your company has a system like that. If they have any supporting materials (like quotes or estimates for the work that needs to be done) that might help the analysis, ask for those too. These can be uploaded to an online document repository to ensure that all the change paperwork is held together.

2. Update the change log.

    The change log is like a risk or issue log and in its simplest form is a document where changes and activities to assess changes are written down. Online tools offer further options such as uploading documents and pushing changes through a predefined workflow, but if you are just starting out, a spreadsheet is as good a way as any to start capturing changes. Appendix 3 has an example. Whatever format your change log takes, document the change there.

3. Assess the priority of the change request.

    Give the change request a priority. Is it critical, important or nice to have? This provides a sense of urgency for planning the impact analysis. Be guided by the person who raised the request but use your own common sense as well. Have clearly defined categories so that one person's 'urgent' is not equivalent to someone else's 'nice to have'.

4. Assess the change.

   As a team, look at the change request. It is helpful to assess all changes against the same criteria. Useful criteria include the impact on:

   - the project schedule;
   - the project budget (remember that the contingency fund is not there to pay for new requirements!);
   - project risks;
   - quality;
   - other project requirements, and;
   - resources.

   The purpose of the assessment is to decide whether to approve or reject the change – in other words, whether to do it or not. You should also consider the impact of the change if it is not done, as sometimes not doing something can have more of an impact than doing something.

   The change request process can be used by stakeholders to bring things to the project team's attention or to raise issues, so you also have the opportunity to check whether it is a legitimate change or a query that is better handled via a different route. If you need to allocate a dedicated owner to the change to assess the impact in more detail, do so. Ask them to report back to you once they have a recommendation. On larger projects, you may have a change control team or forum and if this is the case, the recommendation should go back to them.

5. Decide the course of action: approve or reject the change request.

   Take the decision, and communicate the outcome to the relevant stakeholders and those affected by the change. Update the change log with the outcome and the rationale behind it. Include a note about or a link to where the change assessment can be found so if there is a query or similar change raised later you can find it again easily.

   If the change is approved, amend all the appropriate project documents, like the project initiation document, plan, schedule and budget. Produce and publish a new version of those documents so that everyone can access the most up-to-date version.

Your company's process may be slightly different. Follow internal guidelines if there are any but the activities will be largely the same as described here.

Change management can be unsettling. You start out on a well-defined project and suddenly everything is different and there is a stack of paperwork to complete for the audit trail. Unfortunately, you can't stop changes from happening on projects. 'While it is impossible to stop people from changing their mind about a requirement, it is possible to implement a change request review, validation, or vetting process that protects the specifications from nonfunctional, nonessential changes,' writes Michael Wood in *Project Pain Reliever*.[27] 'This vetting process needs to be comprehensive enough to ensure that requests can be fairly evaluated in terms of the value they add to meeting the project's objectives. Clearly, there is no end of good ideas, but those ideas also have to be evaluated in context to the trade-offs in budget, time, and windows of opportunity that their inclusion into the scope of the project represent.'

The good thing about good ideas is that having them put forward means that people are still interested and involved in your project. They care enough about the outcome to want it to be a success and comprehensive in scope. Unfortunately you can't say yes to every change – otherwise your project will never end and will go drastically over budget and schedule. That's what change control is for.

Once you have been through the change management process for the first time, handling subsequent change requests becomes much easier. Good ideas that are rejected during the change management process can always be packaged up as part of a second delivery phase with a separate business case.

Establish a change management process following the steps outlined here to handle deviations from the original project scope, requirements, schedule or budget.

16 INCLUDE QUALITY PLANNING IN SCOPE

A quality plan is a document that describes how quality is to be achieved during the project. Like any plan, it includes timescales, milestones and resources: namely what quality activities are to take place and who will do them.

BANKING ON PROJECT MONITORING (PART 1)

Ian Duthie, a senior project manager at Lloyds TSB, was working on a human resources project with a vast scope including payroll integration, web-enabled direct access for some HR processes to employees, web-accessed processes for line managers and enhanced management information. The project was part of a larger programme aiming to extend the model of using just one HR department for the entire group – no mean feat for a company operating across over 25 countries, with a raft of subsidiary companies. The project was scheduled to be implemented in a series of releases and involved a team of around 160 people.

On a project of this scale, monitoring the quality of the deliverables was going to be a large job. The quality plan set out the standards for the project and how and when quality deliverables like project health checks and quality assurance reviews would be built in to the overall activities. 'This was a project which had been running under an external project manager,' Ian explains. 'When it was brought in-house we applied our own internal controls to it, one of which was a quality plan.'

The team were fortunate in that they were able to reuse parts of a quality plan from another project. 'We based the plan on something we'd done for another part of the organisation on the same project,' Ian says. To get the quality plan for the £50 million project recognised as part of the official documentation it was essential to have it signed off and approved by the teams involved. 'The overall plan was signed off by the programme manager, while the various deliverables were signed off at the appropriate level within the business and IT departments,' Ian says. 'Some of the approvals took place by way of a "desk check", where the relevant people read through the document. Other approvals were more formal and were reviewed in meetings.'

Having a plan is only half the work: Ian knew that getting the team to stick to it was going to be a challenge. 'We did stick to it!' he says. 'The fact that focus was kept

> on the quality plan was probably due to the nature of the people who were tasked with ensuring that the various deliverables were met.' The project team linked the quality plan to the project deliverable plan and made sure that major points for quality sign off were documented in the main project plan. 'As deliverables were due at various stages throughout the project there was always a focus on the deliverable and milestone plan and the quality deliverables just kept on coming up!' he adds.

The quality plan should include:

- any standards that must be adhered to;
- quality control and quality assurance methods;
- responsibilities: who will carry out the activities;
- quality tools, if you are going to use any, and;
- a reference to the change control process that the project will follow.

The quality plan will also reference the project acceptance criteria and may include the configuration management procedures too.

> Quality control is the day-to-day activity of making sure the work delivered is up to scratch. It is done by the project team.

> Quality assurance is normally carried out by someone outside the project. It is an independent check.

There are two main parts to a quality plan: the definition and standards, and the schedule.

The first part of your quality plan sets out your definition of quality. Quality is a very vague concept, but it does relate back to the golden triangle we saw in the introduction to this section. 'While time and cost are the primary constraints limiting a project team's efforts, many customers primarily judge and remember the perceived quality of the end result,' writes Jeff Furman in *The Project Management Answer Book*.[28] As a result, different customers will each take away a different interpretation of what 'quality' means to them. The only way to understand how stakeholders will judge quality is to ask them.[29]

Aside from stakeholder definitions of quality, you can decide on some quality standards that will apply to your project. Each organisation will have a different definition of what quality means in their environment. This can be a difficult task, so start by thinking: how will we know what we deliver and the way in which we deliver it will be good enough? Start by asking your colleagues or the PMO about what quality standards are appropriate, using this checklist of questions:

- What project management standards or methodologies are we expected to follow?
- What IT coding standards exist in the organisation? For example, how will code be checked and tested before being implemented?
- Are there any industry standards to follow?
- Are there health and safety considerations to meet?
- What other legal requirements or generally accepted norms do we have to deliver to?
- What standard does the customer expect from a final product?
- How will these standards be measured?

Documenting these forms part of your quality plan: you are setting out the standards by which you choose to work. If there aren't any, get your thinking cap on and come up with a framework that your team can agree to work within.

The second major part of your quality plan is a schedule of when quality activities will happen. Some will be integral and ongoing: proof-reading a document before sending it out for comment, for example. There is no need to record that. For each major project deliverable specify:

- what will be tested;
- when it will be tested;
- who will do the testing;
- any tools that will be used, and;
- how the results will be recorded.

Also in this section include any planned dates for quality assurance reviews and who will be responsible for co-ordinating them.

The change control procedure will either be fully documented in the quality plan or readers should be pointed towards another description of how changes will be managed. For more on change control, refer to Chapter 15.

Obviously delivering a quality product at the end of your project, whatever that may be, will go a long way to securing your reputation as a competent project manager. However, you may want to keep your quality activities low-key. Lynn Crawford discovered that paying attention to quality is a characteristic associated by top managers with low performers. Managers in her research saw top performers as people with skills in many other project management areas, but being good at quality management actually decreased the likelihood of being seen as a great project manager.[30] You've been warned.

> Quality is something that must be worked at throughout the life of the project and having a quality plan will define how that can be achieved.

17   WORK OUT HOW TO TRACK BENEFITS

Projects normally aim to deliver some kind of beneficial change. But how do you know if your project has been beneficial?

Success criteria are the standards by which the project will be judged to have been successful in the eyes of the stakeholders. It is these that must be tracked to be able to answer the question of whether your project has delivered any benefits.

BANKING ON PROJECT MONITORING (PART 2)

The human resources project that Ian Duthie was involved with was strategically important to Lloyds TSB, and as one of the objectives was to make the bank's HR processes more efficient and effective for employees it was vital to make sure the benefits were measured to find out if the improvements were making a difference. As a result of the project, line managers and employees would be given direct access to HR information relating to their employment with the bank – a huge challenge for a bank with over 69,000 staff in dozens of countries.

'We had a benefits management tracking system in place and devoted one person full-time to managing it after the main delivery stages of the programme,' Ian explains. As well as bringing all the HR teams together the programme aimed to develop the IT infrastructure underpinning the department's strategy. There were plenty of benefits that needed to be monitored to show that the project had been successful and pick up any areas where there were opportunities for further improvements. 'We wanted to ensure that we tracked benefits accurately,' he adds.

Ian believes that benefit tracking is a part of the project team's responsibility. 'In this project though, the ultimate accountability for the delivery of the benefits remained with the sponsor,' he says. 'The company will make sure the project benefits continue to deliver by writing the achievement of benefits into the targets for the year. So the achievement or otherwise of the benefits will therefore be a prominent feature of the Executives' reward package.'

He has some advice for project managers, based on his 20 years of experience in financial services in the UK and overseas, most of it spent in a project environment. 'Make the decision whether you will devise, adapt or adopt a system for benefit

> tracking and then use it!' he says. 'Think very carefully about the metrics at the outset of the project and be sure of what you are measuring. Consider why you are measuring these metrics: will they actually help you prove whether or not the project has met the objectives set out at the beginning?'

Tracking benefits can be the role of the project manager, but if your project is part of a programme, or your company has a PMO, you may find that someone else takes responsibility for this. In fact, as you deliver the project and then move on to another project, tracking benefits as a project manager can be a problem as most benefits are realised after the project is complete. Instead, a portfolio analyst or business change manager may well work with you to establish the project benefits and this person may take responsibility for tracking them on an ongoing basis. This is great for you – one less thing to do – but it doesn't absolve you from all responsibility.

> Even if it will not be you tracking the project's success, it is a courtesy to the business-as-usual team to put in place a mechanism by which benefits can be tracked – if only so that career-influencing managers will be able to see the successes you helped them deliver six months later.

Terry Cooke-Davies carried out research at more than 70 large organisations to come up with the 12 factors that are critical to a project's success. On the list is: 'the existence of an effective benefits delivery and management process that involves the mutual co-operation of project management and line management functions'.[31] The way that benefits process looks fully depends on the project you are managing and the PMO or company standards that you are expected to follow. However, even if there are no corporate standards, you can still consider how benefits will be managed and tracked on your projects. In fact, on large projects where there are several delivery phases, benefits tracking will often fall to the project manager, as you monitor the success of the previous stages in order to better deliver the later stages.

Whether you are preparing a benefits management process from scratch or supporting someone else who will manage this for you, give some thought to how the project will deliver value to the stakeholders and how this will be measured. Consider the following:

- What benefits metrics can you identify?
- How are you going to measure the metrics you have identified?
- Where will the information come from?
- Is it readily available?
- If not, how much effort will it take to gather useful data?
- Who will collate the benefit measures?
- When is the measurement data going to be available?
- If the different metrics are not available at the same point in time and cover the same period (weekly or monthly), will discrepancies in the data collection cause problems?

Terry White sets out a straightforward two-step approach to a benefits management process in his book *What Business Really Wants from IT*.[32]

1. Identify what you want to measure. If the criteria are complex, like employee morale, break them down into tangible, measurable chunks (this can be easier said than done).
2. Establish the current baseline so that you can track the improvement.

The first step of this approach is the identification phase. While you are brainstorming how you will know if your project has been a success you'll probably come up with success criteria related to the management of the project, which you can refer to in project audits or the post-project review. They will be useful to help focus your mind on the 'business' of project management and relate to doing the project right, completing what you set out to achieve within the defined parameters. However, these are not sufficient alone. Your identification phase should also identify deliverable-based success criteria which are strongly linked to the business case and the rationale behind doing the project. There is no point in having success criteria like 'ensure every project manager takes a full hour break at lunchtime' (unless you are running a project on improving the work–life balance and feel like being dictatorial). Your sponsor and business team have engaged you to do the project to deliver something that will be of value to them and that is what should be tracked. These two types of success criteria are summarised in Table 17.1.

Table 17.1 Types of success criteria

Type of success criteria	Example
Project: things related to the professional job of running the project	Produce and gain sign off for project initiation document
Deliverables: things delivered as a result of the project	Distribute 3,000 educational leaflets to schools in our county

Success criteria can be measured in two ways:

- Discrete: Yes/No
  We did or did not do something
  Examples: Project delivered on time, company gained XYZ accreditation, new branch opened.
- Continuous: measurable on a scale
  We did something to a certain extent, within a target range
  Examples: Improve customer satisfaction scores to between 75 per cent and 100 per cent, increase revenue by 8–10 per cent, rebrand 15–20 offices within Quarter Four.

Project management-related success criteria do not need to be tracked over time and so you do not need to generate a baseline of current performance. Once the project or task is over you should be able to say with certainty whether or not, and to what extent, you met the criteria. The true business benefits, on the other hand, may last for a lot longer. Even a one-off project like changing all the office light bulbs to energy efficient ones has durable benefits. The success criteria could be: 'maintain electricity savings at 40 per cent of previous expenditure for three years.' Tracking the benefits will make certain that the business-as-usual team will be aware when the costs start to increase again – and be able to find out who replaced a dead bulb with a non-efficient one.

The second part of White's process is baselining. A baseline of current performance should be taken before or in the early stages of the project. It should record the current performance against the success criteria **before** the project is delivered. This baseline gives the business-as-usual team something to compare against. It is great knowing that you are now calling back customers within 30 minutes, but if you do not know what the situation was before the project was implemented it is impossible to judge if things are better now. The baseline allows clear identification of performance differences in the post-project world. Be sure to use the same calculations and tracking method to establish the baseline as you plan to do for the ongoing measurement, otherwise you risk comparing apples to pears.

Continuous success criteria always include the possibility of being translated into discrete targets. If customer satisfaction was 82 per cent in March, and the target was 75 per cent, you reached the target. If it was 74 per cent, you didn't. Monitoring benefits on a continuous scale is always better as it allows you to track changes over a period of time. If the customer satisfaction target was reached in April, that's fantastic. But you cannot tell from a yes/no measurement if it was better or worse than the 82 per cent reached in March.

White's model leaves out a couple of steps, and you can see a more complete benefits management process in Figure 17.1. Once you have baselined performance you of course need to deliver the elements of the project that will make a difference to the identified criteria. And then you have to track against these to see if you really have made a difference.

Figure 17.1 Benefits management process

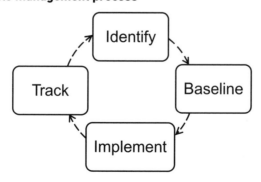

We often talk about baselines in relation to plans, but baseline data – establishing the situation at any given point – can also be used across other parts of the project management life cycle. For example, if your project is supposed to improve something, take a snapshot of the current performance before the project is implemented and then another afterwards. The first snapshot is your baseline; the second allows you to see how much progress has been made.

Despite all the research and good practice about benefits tracking, it is something that very few organisations do well – or do at all. Janice Thomas and Mark Mullaly spent three years working with companies to understand the value of project management and how the benefits from a project management 'division' were managed. Over half the organisations they studied should have been able to show tangible value from their project management efforts, but none of them was actually doing so. These were businesses that provided services to clients, such as consulting and construction and engineering firms. They could have prepared detailed returns on investment statements, but they didn't collect the appropriate before-and-after data and none of them considered benefits tracking a meaningful exercise.

Only a very small handful of organisations could even attempt to quantify the benefits that they had received, and these quantifications were nothing more than anecdotal summaries of assumptions regarding savings and productivity improvements, conclude Thomas and Mullaly.[33]

While it can be difficult to calculate benefits in a tangible way on some projects – and while your company may not care if you do or not – it's good practice to identify success criteria early on in the project and make some effort to work out how you will measure against them. Involving your project stakeholders in the identification process for success criteria and benefits will increase everyone's confidence that the chosen measures are actually the important ones – the ones people will care enough about to make an effort to track.

 Identify the project's success criteria early on and then track against a baseline performance target to measure benefits.

18 ELIMINATE AMBIGUITY

On an IT project you will have probably lived and worked with the details for weeks, if not longer, before handing over the requirements document to the team who will eventually build the system. It is easy to assume that as you know exactly what you mean, they will too – but this is a dangerous assumption to make.

ADDRESSING THE PROBLEM

Sarah, a project manager in the financial services sector, collated the requirements from her business users for a new IT system. They included the need for staff to be able to record clients' address changes. Clients wanted their post to go to their old address until the day they planned to move house, then from that date to their new address. The customer-facing team decided it would be easier to ask the system to store the new address and the date from which it was effective, rather than make a manual note of it and create more work for themselves later.

The system was built with a field for the new address and a field labelled 'effective date'. While it was being tested, Sarah realised that when the 'effective date' arrived, nothing happened. The system did not automatically switch to using the new address. The developers had not correctly understood the need and had not built the functionality the users wanted. Fortunately, as she had planned for an adequate testing and revision stage, Sarah was able to ensure the change to her system was cheap, quick and pain-free – and that in the future, customers would be receiving mail sent to the right address.

Documenting scope is a laborious task, but it is essential to the success of your project that you get it right. That means not only including all the relevant details, but also writing it in a way that is unambiguous. Ursula K. Le Guin gives this advice to aspiring authors in her book *Steering the Craft*:

> Our standards for writing are higher and more formal than for speaking. They have to be, because when we read, we don't have the speaker's voice and expression and intonation to make half-finished sentences and misused words clear. We have only the words. **They** must be clear.[34]

Your project scope document is unlikely to be entered for any literary prizes, but Le Guin's advice is still sound for those of us not writing novels. The project scope is not a document for you; it is a list of detailed instructions to someone else.

On Agile projects, business representatives work alongside developers so scope unfolds as the project progresses. In this environment it is more acceptable to have ambiguous requirements as the collaborative working techniques mean that more clarity comes from working with and continually testing iterations of the end product until it meets the business requirements. On this sort of project you still want to eliminate ambiguity, but it's being done in a different way to an approved scope document, through constant iterative development which refines the requirements in practice during the project.

Non-Agile projects still require a list of requirements prior to commencing work. When you are putting together business requirements for your project:

- Be absolutely certain that there is no ambiguity between what your business users want and your own understanding. Organise a workshop to elicit what it is they really want from your project. This is an opportunity for complete blue-sky brainstorming – if they could have anything as a result of this piece of work, what would it be? Get them to be clear about their requirements, ask the stupid questions, then negotiate. It might never be possible to have a dedicated customer service representative for each client or response times to customer calls within two seconds, but those requirements allow you to form an understanding of what is important to your customer: in this example, the business user.
- Once you have the list, add to it making sure you are being as specific as possible. List colours, materials, brand constraints and any legislation with which the solution must comply. Are there any other systems with which it must integrate? What security is appropriate? Detail exactly what is meant and spell out any assumptions.
- Verify the document with the project customer before it goes to anyone else. In other words, check that you have understood and documented their requirements in a way that is meaningful and clear. Include wireframes, screen mock-ups or models if this helps eliminate ambiguity further. Once the customer is happy that their requirements have been adequately captured, you are ready to start the next stage of the project and build the solution.

Allowing the project stakeholders to review the document also means that if the system does not deliver what they were expecting when they first see it you are in a better position to explain why: you gave them plenty of opportunities to add new requirements or clarify existing requirements. If the stakeholders didn't take those opportunities they will have to use the formal change control procedure to make any alterations or add new requirements to the project scope at this stage. This is where Agile project teams have a huge advantage over traditional, waterfall ways of working. They will quickly pick up changes to requirements and be able to adapt to them – consider how you can continually involve your stakeholders in the project to ensure that as the design and build progresses that they are getting what they want.

This assumes that you'll be preparing the requirements yourself, but if you can draw on a skilled business analyst, they will probably be better than you at doing this. Business analysis is a toolset aimed at eliciting requirements and ensuring solutions are fit for purpose, among other things. Business analysts are excellent facilitators with a good working knowledge of business processes. They can work alongside you on a project to eliminate ambiguity and provide the 'translation' from business-speak to project-speak that you and the rest of the team may need.

A good requirements document (whether it is written by you or a business analyst) should not only ensure your product is built exactly as you wanted, but also forms a way of checking off users' needs at the testing stage. It will form the basis of the user testing documentation and if it is very detailed, for example including process maps, it may save you time in writing test scripts.

> **I DON'T HAVE TIME TO DO ALL THIS!**
>
> It is inevitable that at some point in your career you will work on a project where there is simply not the luxury of time for a full requirements document, even if you do have someone to help. Work with your team to revise things as you go along. It can be risky, but if it is a well-thought-out and relatively simple project, this approach can save a lot of time. Just remember to update the requirements document as you go along so at the end you have an accurate record of what has been done. If you do opt for the make-it-up-as-we-go-along approach, make sure you allow enough time in the plan to test your solution thoroughly, both technically and from a user's perspective. It is much easier to alter things at the testing phase than after the project moves into a full implementation.

Finally, if you are in any doubt about whether or not your team can handle the ambiguity, take the time to properly elicit requirements and to document the output fully – it really will be worth it.

> The users' requirements form the basis and rationale for the entire project, so it is essential to have them:
>
> - documented;
> - unambiguous, and;
> - agreed by the users themselves.

19 USE VERSION CONTROL

Version control is a method 'of recording and monitoring the changes made to a document over time'.[35] It is a simple way of being sure that everyone is reading the same edition of a document and that everyone has the same view of events. It **is** a discipline but it is not difficult to do, especially if you take advantage of the in-built version control functionality in many software applications.

THE RIGHT VERSION OF EVENTS

Version control is very important for project documents, according to Graham Inglis, an experienced IT project management consultant, who lives and works in France. 'It's essential to be able to identify the most recent version of pretty much any project document, and also to be able to identify the version used in connection with historical decision points,' he says.

There are manual and electronic ways to perform version control, and Graham has used them both, although prefers electronic methods. Version control becomes particularly worthwhile when you need to refer back to something. 'I have had to draw on a previous version of a document to establish whether a particular technical restriction had been identified at the time of a particular change request,' says Graham. 'The customer said it had; I said it hadn't.'

Graham advises that all plan documents should be under version control, as should all technical specifications, and anything else that captures decisions at a point in time. 'Documents such as best practice sharing do not really need to be under version control as long as the latest copy is easily identifiable,' he adds.

You can put any documents – and pretty much anything including project deliverables and IT code – under version control. For documentation, you don't need a special database or filing system, but an electronic document repository with built-in versioning capability will make things a lot easier.

A low tech way to do version control is simply to note the details of the version within the document itself. Do this on the front page of the document, or in a footer.

The easiest way to capture the whole document history is in a table with the following columns:

- the version number;
- the date of this version;
- the author(s), and;
- the reason for this version and/or a summary of any modifications made.

The numbering is really important and is the key to successful version control. The numbers will be used by your team to check that they have the same version as everyone else. Here are some numbering rules:

- All drafts are written as version 0.1, 0.2 and so on.
- If the document requires sign off, a copy of the last 0.something version is signed (either physically or electronically, or if your company is relaxed about this sort of governance, verbal approval may be enough). Then the document is reissued as version 1.0 to all appropriate parties.
- If the document does not require sign off, the version which will be circulated as the 'official' copy is made version 1.0.
- Small changes that result in new versions become 1.1, 1.2 and so on.
- A large rewrite or new round of signatures for a significant re-approval becomes version 2.0.

Have a look at how it works in practice in Table 19.1 and read through the hints in Table 19.2.

Table 19.1 Version control table in a sample document

Version Number	Revision date	Author	Summary of changes	Changes marked
0.1	1 February 2013	Nanette Bailey	First draft	No
0.2	15 February 2013	Nanette Bailey	Reviewed by project team; section 6 updated (new dates)	No
0.3	17 February 2013	F Jacobs	Final review by all stakeholders; costs updated	Yes
1.0	20 February 2013	F Jacobs	Issued	No

Most collaborative working project management tools have the ability to manage version control for you. Microsoft SharePoint has a sophisticated take on version control, allowing check out of documents so that only one person can work on them at the same time. When a document is checked in again, it is given a new version number. Previous versions and a summary of changes can also be seen, so the whole document history is there in case you need to go back to a previous version to check something.

Other products have similar capability, allowing users to store multiple copies of documents and making sure that the current version is always the one available to users which avoids anyone downloading or reading a previous version by mistake. Check what versioning capability your tools have and use it – it will save you a lot of time.

If version control is already in use in your organisation or on the project but using a different logic to that described above, then stick to that. Don't get yourself and the team confused by trying to adopt a different standard. The important thing is that you all know what your chosen methodology is, how it works and that you use it without fail.

Table 19.2 Tips for version control

Do...	Don't...
• Make sure your collaboration or project management software has version control switched on if this setting is available	• Forget to update version numbers in document footers and headers
• Tell everyone on the circulation list when a new issue is released and tell them where they can find a copy (or ensure they are alerted through your collaboration tool settings)	• Issue new documents for every changed comma. If the modifications are of little importance, save them up for the next release
• If this document is referred to in other documents, update those references to reflect the latest version of this document	• Throw away paper copies with hand-written changes on until you are sure they are no longer required
• If you are manually tracking versions, keep copies of previous versions. Set up a separate folder called 'Archive versions'	• Hesitate to ask contractors to use the same version control methodology as your team
• Get the right people to approve and review documents first time round to avoid going around the sign-off loop again and issuing another version	• Try to use version control for a document that changes daily like a risk log
• Explain to everyone how version control works and why they can't just change documents without you knowing about it and managing it in a controlled way	• Forget to do it!

Version control uses minimal effort (especially if you tap into the functionality of your collaboration tools) and will guarantee that you always know which is the latest copy of a document.

20 PUT A POST-PROJECT REVIEW IN SCOPE

A post-project review (PPR) is a debrief at the end of the project which analyses what went well and what were the key challenges. The objectives of a review like this are:

- to bring everyone together at the end of the project to formally close it;
- to formalise the key lessons learnt during the project, and;
- to record this knowledge in such a way that it can be used by other projects to avoid the mistakes your project made or to benefit from implementing things that worked especially well.

Many organisations do not routinely carry out PPRs and have no formal way of capturing and sharing project learning. If this is the case in your organisation, it is still very worthwhile putting a PPR in the scope of your project. You and your team can personally benefit both from highlighting the pitfalls encountered in this project to avoid in your next endeavour and for the closure such a review brings.

AN ALTERNATIVE WAY TO RECORD LESSONS LEARNT

While it is a good thing to include a PPR at the end of your project, an alternative and additional way of capturing lessons learnt is to make it an ongoing part of the project management process. During a project to deploy a new diagnostic imaging software product across a number of hospitals, a UK healthcare firm chose to do mini-reviews with project stakeholders each month, including the project team members themselves. The project manager asked questions about what was working, what wasn't, and how the stakeholders felt the project was going. This allowed future implementations at hospitals later in the deployment schedule to benefit from the lessons that had been learnt at earlier sites. Communication, proactivity, delivery timescales and overall satisfaction with how the project was progressing were all assessed, and the scores collated monthly to graphically track customer satisfaction with the project. At the end of the project, the whole team could see how the project had evolved and had confidence that what was identified early on was put into practice during the project. In this way the team benefited from continuous improvement, instead of waiting until the end of the project to review what worked and what didn't – by which point it would have been too late to do anything about it.[36]

Including a PPR in the scope of the project will be a step towards making sure the review actually gets done. Add the PPR into your scope statement and the report from the PPR meeting as your project's final deliverable. As the project moves into its final stages, book the date and a room for the meeting and start asking people to keep the time free. Bear in mind though that the fundamental reason for doing a PPR is to ensure that the resulting information is shared and added to your organisation's corporate knowledge base, to avoid 'project amnesia.'[37] This can be done in a very formal, structured way through a database or wiki of lessons learnt, for example, or in an informal way where you help the team to reflect on the key messages so that as individuals they will have benefited from the experience of the project and take that forward into their next assignments. In reality, you will probably end up being some way along the spectrum of possibilities, producing a report of successes and challenges which will be circulated to the project team and interested parties.

USING A WIKI TO CAPTURE LESSONS LEARNT

A wiki is one way to capture information and lessons learnt along the way during a project. There are a number of wiki software packages, many of which are free, so it is relatively easy to get started using a programme to capture lessons learnt as you go. This could either be at a project level or a PMO level, in which case you would structure the wiki differently to make it searchable by category, project name or department that the lesson applies to.

Make sure that everyone has access to update it, so that lessons can be recorded in real time. You could also set up alerts so that every time a new lesson is added, team members receive an email notification prompting them to look. The best way to manage access is through named users so that you avoid anonymous access which will prevent you from seeing who made changes and could potentially lead to abuse.

At the end of the project you can refer to the lessons added to the wiki – this can be a good way of helping people remember what they learnt as on long projects it can be difficult to recall information from the early days. You can also hand the wiki entries over to the operational team at the end of the project as part of the closure activities so that they too benefit from the information you learnt during the project.

The PPR should be held at the end of the project and involve as many of the people who contributed as possible. Invest some effort in thinking about what kind of format you want for the review. A meeting where you go through a checklist of questions relating to each phase of the project (see for example Table 20.1) may not prove to be the open and honest learning environment you were hoping for, especially if the project hit some serious difficulties. You might decide to interview each individual or team separately, collate and circulate their responses anonymously, then hold a brief review meeting with everyone in order to agree and share the output. Whatever the format, the same questions will need to be addressed:

- What are you evaluating? Which parts of the project fall into the scope of the PPR? If some parts are excluded, why? Will they be reviewed separately?

- What criteria will the project be assessed against? Can you refer back to the project's stated objectives and success criteria?
- Will you do the PPR yourself? Sometimes it can help to bring in another project manager or an experienced facilitator to run the meeting so you can fully participate. It also makes the environment more neutral, which is especially helpful if you expect sparks to fly between team members.
- What format will the output be in? Are you aiming to produce a report, a memo, a set of database entries or more data for the wiki? Knowing this will help you structure the meeting to get the best output for you, and potentially the operational team who will be managing the outputs now your role on the project has come to an end.
- What ground rules do you want to set for the session? It can be useful to specify at the outset that the meeting is not aiming to apportion blame for failures. Agreeing some ground rules can also help (turn off mobile phones, constructive criticism only, avoid commenting about an individual's performance: depersonalise feedback with language relating to their role in the project or their department and so on). Pin up the ground rules at the beginning of the meeting so they are available to be referred to as the session goes on.
- Consider getting agreement on whether each point can be documented. Be sensitive about what remarks may best be kept within the confines of your meeting room if the PPR report is going to senior management.

Table 20.1 PPR checklist: example questions to ask during the post-project review meeting

Initiation/Start-up	• Were the project objectives, scope and critical success factors identified and agreed? By whom? • Was a project organisation established, with clearly defined roles and responsibilities? Were these documented and signed off? • Was a cost–benefit analysis or business case drawn up and signed off?
Planning	• Was the content of agreed deliverables established? • Were all products and responsibilities agreed? • Were dependencies recognised and key milestones identified? • Was a meaningful schedule created? • Was a critical path established?
Monitoring	• Was progress monitored against the plan? • Were issues, risks and dependencies recognised and managed in an appropriate way? • Did the project schedule and budget prove to be realistic and achievable?

(Continued)

Table 20.1 (Continued)

	• Was effective action taken to address: changes to scope, poor quality, time slippage, cost overages, resource issues? How was this done? Did it work? • Were changes to the schedule and business case appropriately approved and communicated? • What status reporting was done and was it accurate?
Structure/Organisation	• Was the established project organisation effective? • Did the project team function well, sharing views and opinions and owning joint decisions? • Did the steering committee or project board function effectively? • Was communication effective at all levels? If not, why?
Outcome/Result	• Was the project delivered on time and within budget? • Is the quality of the solution robust enough for the job? • Does the delivered solution meet the business objectives and detailed requirements? • Has an operational handover already been done? If not, when is it scheduled for? • What are the outstanding tasks and who will take responsibility for them?

Plan for a post-project review at the beginning of your project but don't forget to review lessons learnt and tweak your approach along the way. Put the PPR in scope and then follow it through, sharing the output with the team and organisation for other people to benefit.

21 IDENTIFY RISKS UPFRONT

Successful risk management starts at the beginning of a project with the identification of risks: potential happenings that may throw a spanner in the works of your project. This chapter and the following two look at the risk and issue management process. This chapter considers the risk identification phase done at the beginning of a project. Chapter 22 continues the risk management process by discussing risk response and management. Chapter 23 explains the difference between risks and issues and how issues can also be managed effectively.

DRIVING STRAIGHT IN

The life cycle of UK government Ministry of Defence projects includes an upfront Assessment Phase before time, cost and performance targets are agreed and the project is formally approved. The purpose of the Assessment Phase is to identify and understand the critical risks and put plans in place to manage and mitigate against them. The Ministry's recommendation, based on their last 40 years of experience, is that about 15 per cent of the initial procurement costs should be spent on this phase.

The Support Vehicle project is a procurement initiative to replace the current fleet of ageing cargo vehicles with new cargo and recovery vehicles and recovery trailers.[38] In March 2001 it was decided to bypass the Assessment Phase and approve the entire project straight off. The decision was taken because the team believed that the technology was well established and there was already enough information about the project. However, skipping this phase meant that critical risks were not identified and the project was significantly delayed. 'Nineteen months of the delay to the Support Vehicle project are directly attributable to the decision to bypass the Assessment Phase,' says the Ministry of Defence's *Major Projects Report*.[39] As a result the project missed the opportunity to examine risks and plan mitigating actions early on. The tasks had to be done after the project had been given formal agreement to proceed.

Identification means taking the time to note down everything that might go wrong, from key resources moving off the project, changes to the legislative framework, system testing taking longer than planned, changes to internal priorities, any concerns about using new technology and so on. The project manager alone cannot identify all the potential risks, so involving the project team will help generate a comprehensive list.

Elkington and Smallman have studied the risk management process from an academic perspective and arrived at this set of guidelines for risk identification:[40]

- Identify the obvious risks first.
- Think of the who, why, what or when of the project and identify risks relating to those.
- Consider risks that apply to the management of a project as opposed to the deliverable of your project, like resources.
- Identify positive as well as negative risks, for example the impact of one task being completed much earlier than expected.[41]
- Use your imagination to cover everything: removing risks from the list is much easier than managing risks you never identified.
- Involve others by working in small groups or holding informal interviews.

Once you have identified as many risks as possible the next stage is to apply some sense of priority assessment to each item. You will have a very varied list of risks, some of which will seem small and pointless, others very significant. There are two attributes that should be considered for each risk: likelihood and impact.

- Likelihood: the chance of the risk occurring. Unless it is a very scientific event you'll have to take a best guess based on your gut feel about this.
- Impact: how serious the outcome of the risk would be if it materialised. The impact could be to the project schedule, budget or the quality of deliverables, and again you will probably find yourself being less than scientific when it comes to assessing this.

The two attributes are given a value and these are multiplied together to calculate the overall risk assessment.

Figure 21.1 shows a risk matrix and two examples of risk assessment in practice. The first example is the lack of resources for user testing. There's a possibility that users would not be available to carry out testing when needed as they all have operational priorities. This has been assessed as a 4 on the risk likelihood scale. The impact on the project timescales would be a moderate 3.

The likelihood of facing an office flood is tiny. This has been assessed as 1 on the matrix. The impact on the project if the office was flooded out would be severe, and that has been rated as a 5.

Combining the likelihood and impact scores gives each risk a relative priority. User availability for testing (4 x 3 = 12) is a more significant risk than flooding (1x5=5). Once each risk on your list has been assessed and given a score based on its priority you will be in a position to consider how best to manage it, which is discussed in the next chapter.

Figure 21.1 Risk matrix

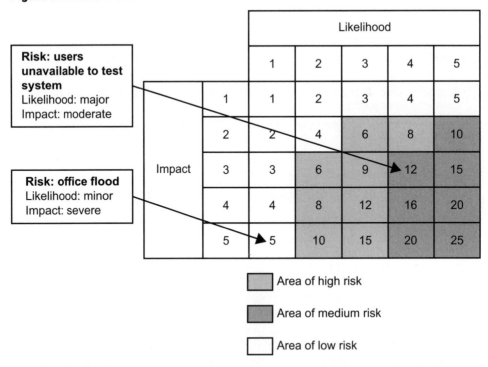

 Identification of risks should be carried out early in the project involving the project team and each risk should be assessed according to impact and likelihood.

22 MANAGE RISKS

Having a detailed list of risks at the beginning of the project becomes a purely academic exercise if you do nothing to manage them. Ongoing risk management should be built into the project management tasks you do on a regular basis, keeping you on top of anything that may upset the successful delivery of the project.

MISSION CRITICAL (PART 1)

Dr Ady James, a project manager with 14 years' experience in the management of space projects, led an international team of around 30 people to develop a highly technical extreme-ultraviolet imaging spectrometer. The instrument forms part of a Japanese spacecraft which has been designed to study the sun.

The project started in 1998 as a collaborative effort between the UK, US, Norway and Japan, with a budget of around £13 million. Managing the risks associated with such a large and critical piece of work was essential.

'The main aspect of our risk management system is that it is paper-based and under the control of the project manager,' Ady explains. 'It requires very little maintenance other than the vigilance of the project manager.' Deceptively simple, the project team was constructed in such a way as to make risk management foolproof. Ady's team, responsible for the spectrometer, tracked risks relating only to that. 'The spacecraft-level team had their own risk assessment and so we would be tracking the same risks. Similarly the local teams had local management and would be tracking their risks. My risk identification therefore passed some risks up and some down and I received risks from both directions. This acts as a failsafe for spacecraft-level risks. If we didn't identify it as a risk then someone else might.'

Near the start of the project the whole project team were asked to identify potential risks. These were then added to a draft risk assessment document. A small group of the project team with responsibility for the delivery of the technical elements of the project reviewed the document and categorised the risks. Risks were divided into two major sections to help with the classification:

- Programmatic risks, associated with the build phase of the instrument up to launch.
- Operational risks, associated with the operations phase of the mission, i.e. post launch.

Then they were further split down into another two sections:

- System level risks, i.e. these are defined as those which have impact beyond the experiment and may affect the spacecraft or mission.
- Sub-system level risks, i.e. those which are contained within the experiment and are therefore only likely to impact performance.

'This is where we split away from current thinking on risk management,' Ady says. 'Normally one would look at probability of occurrence and the impact if the risk was realised. Generally any impact on performance will be unacceptable to the end-user, the scientist.'

There were performance tolerances for the spectrometer and the role of the risk management exercise was to resolve any risk that put the instrument outside of this tolerance. For that reason the risk impact was documented but not scored – any impact was treated in the same way.

'Another split from traditional thinking is that, rather than an individual, we name the local manager as the owner of the risk as someone who can locally delegate,' Ady says. 'It is up to the local project managers to report to me on risk and assign an owner internally if appropriate.'

The project team updated the risk log and reviewed the list of actions during their regular meetings and at the end of significant phases the log was reviewed by a panel. Occasionally new risks were added at this point, and the panel had to be satisfied that all risks were being adequately managed before allowing the team to move on to the next phase.

Ady had originally planned to update the log with new risks on a monthly basis, but in practice there were not that many to add. They tended to be minor performance concerns or relating to unplanned work. 'I found that the frequency of controlling all of the management tools was driven more by the needs of the project than by any active plan on my part,' he says. 'When things were going as well as expected a dogmatic adherence to updating risk logs, actions list and chasing the individuals for the sake of an update did not seem productive.' Ady believes project managers can get a feel for this from 'management by walking around'. 'You get a feel for the team's concerns and know when you may want to formalise some of the control functions,' he explains. 'As the team matures into the project – many of mine were already experienced – I found that they were all so aware of the needs of the project that management was always more of an oversight activity rather than a forced driving-type activity. I have the advantage that nearly all of the team members are suitably motivated by the very nature of the work but I suspect that not many industries get that for free.'

The majority of the risks related to the fact that the project was using untested technology. 'The number of risks in this type of project reaches a plateau very early on and additions are rare,' Ady says. 'In our domain we find that risk management rapidly integrates with the project scope, rather than being seen as a separate

> activity, so updating the risk log is not so essential. The level of risks at the sub-system level is also generally low as the result of a fairly conservative industry and an experienced development team.'
>
> Ady kept the log tidy by closing down risks that had passed and not materialised, which were mainly those relating to late deliveries or test failures. 'Despite the low number of issued updates to the document it was at least read by me on about a monthly basis,' he says. 'I set up reminders on my calendar software to remind me to check and update the risks, as well as the budgets, schedules, actions and issues, on a monthly basis.' Ady's advice to other project managers is to never plan to do these updates on a Friday. 'Don't plan to do them all on the same day either,' he adds, 'because it won't happen. You never get a full free day when a project is in full flow.'

Risk management can be done in an incredibly formal way, with scheduled meetings where the team gets together to discuss the latest developments for each risk, or in a less structured way, with the project manager co-ordinating the management activity and getting updates on risks as part of your normal progress updates on the team. How you choose to record your risks is up to you; a starter-for-ten document template is included in Appendix 1.[42] Many online project management software solutions include risk management databases or forms and the advantage of these is that everyone has access to the current version and latest status at all times. However you choose to record the risks, the point of having done the identification exercise is to now put plans in place so they never happen.

> Not all risks will have a negative impact on the project. Some risks may have a positive impact and you will want to encourage those ones to materialise. For example:
>
> - Our call centre may not be able to deal with receiving so many phone calls about our new product.
> - Great weather may encourage more people than expected to arrive at our open day.
> - Our new website may receive more traffic than we anticipated.
>
> All of these risks have a positive effect on the business through generating more interest and revenue than you had perhaps anticipated. It is worth planning for positive risk as well (often called opportunity), and to do the most you can to encourage these risks to happen – capitalise on the positive opportunities while trying to minimise the negative risks!

Whatever your approach to the paperwork side of things, your approach at the 'business end' of risk management is going to be the same: work out how to handle each risk, plan actions to meet that strategy and monitor progress against the actions. This is the actual 'activity' relating to risk management as opposed to the documentation and thinking process that has to happen upfront. The activity of risk management is the critical part as studies have shown that having an up-to-date risk management plan and a process for assigning ownership of risk statistically improves your chance of completing the project on time.[43] The more risk management you do, the more chance you have of a successful project delivery (however your project defines success). It is hardly surprising that research shows that the riskier the project the less successful the outcome; some projects, however much risk management you do, however many times you allocate actions or chase your team updates, are just too risky to really deliver a top result.[44]

The next step after risk identification is to work out what you are going to do about the risk. This is called risk response. There are four risk response strategies for negative risk, as shown in Table 22.1, and four strategies for managing positive risk, as shown in Table 22.2.

Table 22.1 Risk responses for negative risk

Response	Description	Example response Risk: bad weather may spoil the school fete
Avoidance	Refrain from carrying out the activity that will result in the risk occurring	Don't hold the fete at all
Reduction	Act to reduce the impact of the risk should it occur or the likelihood of it occurring	Hire marquees for all the stands so the fete can go ahead under cover if the weather is bad
Transference	Get someone else to take on part or all of the risk	Take out an insurance policy against the potential loss of income for the school if bad weather keeps people away
Acceptance	Understand and accept the consequences should the risk happen	Accept that there is a chance of bad weather and do nothing

Baccarini, Salm and Love have looked at of the popularity of each of these different types of response to negative risk in a study of Australian IT project managers. Presented with a list of frequently occurring risks that were not specific to any particular project, like lack of resources, incomplete requirements, unreasonable project schedule and so on, the project managers described the strategy they would apply to managing the risk.

Table 22.2 Risk responses for positive risk

Response	Description	Example response Risk: more people than expected attend the school fete
Enhance	Act to increase the probability of the opportunity	Promote the fact that the weather is due to be good on all the posters; reduce the entrance fee
Exploit	Act to eliminate the uncertainty so the opportunity definitely happens	Unable to do in this case but good publicity, good quality stands and parental support will help; increase the number of helpers and facilities so that if it happens the school is prepared
Share	Get someone else to take on part or all of the opportunity	Set up a team of parents and volunteers to promote the fete and drum up support; invite another school to take part as well
Acceptance	Understand and be open to the consequences should the opportunity happen	Accept that the fete might be crowded but do nothing

The results were incredibly varied. The authors conclude that this indicates that there is not one solution for managing any particular risk and the project manager must be aware of the possible need to implement two or more treatments for one risk.[45]

As their research shows, there is no textbook way to manage any given risk. What works for one project in one situation may not work for exactly the same risk but a different project and a different situation. You will need to use your judgement to decide what action plan will be the most effective for you. Baccarini, Salm and Love's analysis does show that reduction is the most favoured risk response for negative risk. Transfer and acceptance are hardly used at all. This may be because in the majority of projects it is difficult if not impossible to transfer the risk to a third-party contractor, an insurance company or even another department. Acceptance is similarly not frequently proposed as a risk response because it is only suitable for very small risks. It is rarely appropriate to do nothing.

Your risk management plans can become a formal part of the sign-off process to move from one phase of the project to another, but it's not a good idea to only talk to your sponsor about the project risks at this time. He or she will probably want greater visibility of the risks facing the project, so you should provide that information on a more regular basis. One way to do this is to include the top three risks in your regular reports to the project board, along with a brief statement of how you are managing them.

A final thought on risk management: once you have successfully mitigated against a risk to the point where the risk will now not happen (or the opportunity has passed), it can be removed from the risk log. That does not mean deleting it from the document or database altogether. Each risk should have a status: open or closed, so simply change the status of the risk to closed and ensure it doesn't print or display on reports unless you specifically want it to. The record of how you managed the risk will be useful for your post-project review, for an audit or if a similar risk comes up later in the project.

Having a list of risks is not enough. Risks should be managed by defining the appropriate risk response, allocating an owner and carrying out activities to actively manage the risk.

23 MANAGE ISSUES

An issue is a problem that is affecting your ability to deliver the project successfully. It might be big or small, something that can be fixed in a day or approached with a long-term vision, but all issues will be handled in the same way.

MISSION CRITICAL (PART 2)

Dr Ady James's spectrometer project did not have an issue management system in place at the outset in 1998. He adopted an informal approach to documenting issues, combined with a close knowledge of his team's activities and concerns to ensure they were managed effectively.

'The issues log was informal and anything could go in it,' he says. 'I would cut and paste sentences from emails from concerned engineers and give it an issue number so it could be tracked.' Ady led his international team from the Mullard Space Science Laboratory at University College London and engineers from the UK, United States, Norway and Japan contributed to the issue list. Ady noticed that the issues raised fell into two categories: everyday concerns and insecurities, and new items that had not been conceived in the project planning. 'The issues list was a way of allowing anybody to input new tasks into the project plan,' he says. 'Unplanned work issues or points relating to practical build or assembly were raised as issues in the issues log where they were turned into new work packages and therefore planned tasks or actions.'

With a budget of around £13 million, partly funded by the Particle Physics and Astrophysics Research Council, it was essential that every concern was addressed. The issues log was one of the ways Ady could demonstrate to his team and stakeholders that things were being managed. Like the way the project managed its risks, the system was very simple and based on the circulation of documents rather than any complex IT package. Ady is clear that expensive issue management applications are not necessary: 'I like the paper-based systems in that I only have to worry about the update and maintenance of the information and not the update and maintenance of the system that contains the information as well,' he explains. 'Your project tools should support your project and no more. There is a fashion for believing that the quality of the project is reflected in the quality or complexity or newness of the tools used or in how the information is displayed. I don't believe this. Some tools will be necessary when the size of the project is such that a paper

> system is unworkable but if it is not needed don't use it, the management of the project should be challenging enough as it is. When there are major problems teams like to see that the right decisions are being made, actions allocated, resources freed up. The use of these tools, whether paper-based or an IT system, then becomes a good way to show how we would get over the issue,' he says.

Issue management requires the same approach as risk management: log the issue, devise an action plan, carry it out and monitor the situation. The issue log should include any problems that occur which will have an impact on your ability to deliver the project successfully. An example log is included in Appendix 2.

Issues may be the result of a risk that has materialised but could just as easily be something that has never been on your risk log. The log can include issues over which you have no direct control, as well as those you can fix using the resources within your project organisation. However, this is not the place to log approved changes. They should be kept separate so if the result of an issue is to raise a change to address something, the change should always go through the change management process.

By their nature, issues are more immediate than risks and as such you may choose a more informal approach to documenting them. As the situation has already happened it is less useful or necessary to craft a descriptive paragraph explaining the issue: a cut and paste from an email may well be sufficient to provide enough of a memory-jog so that your team knows exactly what is being referred to. Another way to approach documenting the issue is to record the discussion either in video or audio format so that you have a record of the context of the issue and the impact. Your issue log can then refer to the recording.

The appropriate resolution for an issue may be immediately obvious and you'll quickly see what you need to do to create an action plan to rectify the situation. That's rare though. It is more likely that the full extent of the problem will not be known, in which case you and the team will investigate the issue thoroughly in order to be able to find an appropriate solution.

Due to the immediacy of issues, there are two key things to bear in mind:

- the efficacy of the action plan to handle the problem, and;
- the speed with which you allocate someone as the owner of that issue to take responsibility for seeing the action plan is carried out.

Even issues outside of the project's control can have a team member allocated a watching brief, someone tasked to provide updates on a regular basis as an issue unfolds. It may take a significant time for some issues to either be resolved or to disappear. On the other hand, some issues may be a short sharp shock to your project, over and done with very quickly.

MANAGE ISSUES

Issues can be controlled in various ways depending on the type of problem, but a common way to resolve simple issues is to propose a change to the project schedule, budget or requirements to incorporate the tasks needed to resolve the problem. If the project sponsor agrees, the change is approved and you can replan your project accordingly, proceeding with the new status quo.

The issue status in the log can be changed to 'closed' once an issue is controlled. As with risks, do not delete it from the log. The complete list of issues with the action plans and final resolution provides a useful audit trail and input into the post-project review. It will also help you remember what you did if you come across a similar problem later.

Issues have already happened, so log them, draw up an action plan to manage them and move quickly to execute that plan.

24 DOCUMENT ASSUMPTIONS

In any project there are things you don't know. It is impractical to wait until everything is a known fact before work on the project starts. In order to start work, it is necessary to form some assumptions – statements about what you believe to be the case – to create a position from which to begin. These will be proved or disproved as you work on the project.

WORKING IN THE DARK

'One killer in terms of scope and budget for projects can be assumptions,' says Neville Turbit, a convenor of the Australian Computer Society Project Management group and principal of the project management software and consultancy firm Project Perfect.

He spent some time working with an Australian government department to help them implement a project management methodology, which included handling assumptions in a pragmatic and practical way. Teams often find themselves in the dark at the beginning of a project which means assumptions form a necessary part of the foundations of the requirements and plans. 'I talk to my teams about travelling down a tunnel with only a flashlight. We can see clearly what is immediately in front of us, but only some of what is down the tunnel,' Neville says. 'We know there is more down the tunnel than we can see. We just don't know what it is. That's the way it is with scope. We know there is more scope than we can see. We just don't know what it is until we get further down the project tunnel.'

It is the same with assumptions about budget and time. 'At first I started trying to get them to add a contingency to the budget. And I tried to get them to leave a hole in the Gantt chart for the unforeseen. There was so much resistance to unallocated resources that this proved almost impossible. The line management way of thinking does not fit with a project environment and its so many unknowns.'

The government department learnt to handle assumptions as ongoing open issues, with actions required to validate them. Neville gives an example: 'If we make an assumption that we will not have a distributed database, then the project team immediately creates an action to validate the fact that it will not be a distributed database. The action is assigned to a person, and given a date for completion. That action is then monitored by the project manager.' This means that assumptions don't stay assumptions for very long. Either the result of the investigation is that the assumption is true and it becomes a known fact which aids the further development of the project. Or it is false. 'If it proves an incorrect assumption,' Neville says, 'we call a meeting to look at its impact and what other actions may be required to compensate. Too often in projects I have seen assumptions documented and forgotten,' he says.

Assumptions can be:

1. things that you are taking for granted will stay the same;
2. things you have to assume because you don't yet know for sure.

It is worth documenting the first type of assumption in your project initiation document or as part of your project records database even if everyone accepts that is the way things are. Just because something is like that now does not mean it will stay that way. The way organisations work changes more often than a project manager can keep track of: a change in an area where you least expect it can, and will, interrupt your plans. Examples of this type of assumption are:

- Payment for contractors will stay at £900 per day.
- The marketing department will make resources available for testing.
- The April version of the IT security standards will be used to develop the application.

If payment rates change, marketing cannot provide testers or the security standards are updated, you suddenly face a very different project – one which you probably will not have the budget or time to complete. You will still be the project manager and be expected to deliver but having documented the assumptions it will be easier to explain to your sponsor why you now need more money, time or people. They originally agreed to a project in a certain environment: now that has changed, with all the relevant knock-on impacts to the project budget, scope and timescale.

The second type of assumption will allow you to plan more accurately. You might not yet know the number of staff who will need to be trained on the new accounting system, but you can assume it will be half the finance department, say 35 people. Having stated and documented this, you can decide that they will be trained in two groups and book those dates, plan the costs for the trainer, room hire and so on. Planning based on assumptions almost always results in a potential issue, so add an issue to your log: 'Number of delegates to be trained still undefined.' Allocate an owner to the issue and make them responsible for validating the number of staff who require training. Once you have a concrete answer, you can amend the plans as appropriate.

Whether you use database software, a spreadsheet, a formal project management information system or the logging system mandated by your PMO, make sure that your assumptions are recorded in a way that shows the links between them if you can. Often the software used to record risks, issues, changes and assumptions displays information in tabular format and it can be difficult to see the associations between them. Mind mapping software gives you the option of showing links in a more fluid format and may be more useful, especially in the early stages of project planning.

A word of warning: use assumptions with care. It is always better to have the full picture and work with concrete facts. Making assumptions like 'I will have a budget of £500,000 to set up the company football team' will not help you manage the project at all. The fewer assumptions you have, the more likely it is your project will avoid surprises later on and the more realistic your plan will be.

Document all project assumptions in the project initiation document. Document and monitor the associated issues and update your plan when you have validated each assumption.

25 INVOLVE USERS IN SCOPE DEFINITION

Project communication involves both communicating to (or with) stakeholders and receiving information from them. This chapter looks at one of the methods to elicit information from your team and relevant parties. The following chapter considers how to give information to those key stakeholders. Involving users in defining the project scope is fundamental to the success of any project. You can't define the scope sufficiently alone, by brainstorming what you think the users really want the project to cover as you run the risk of both alienating your users and missing something important. Including your customers, the end users, in this part of the project definition will guarantee your scope is as comprehensive as possible. Your role is to elicit all the necessary information from them and guide them towards defining a useful statement of project scope.

FOOTBALL FACILITIES FOR THE FUTURE

Built on 330 acres of parkland near Burton upon Trent in the UK's National Forest, St George's Park is the new training base for all 24 England football teams. The project to launch the new £105 million facility was project and cost managed by programme management and construction consultancy Turner & Townsend. The site includes state-of-the-art training and coaching facilities, a sports medicine centre and two hotels – and it was delivered on budget and completed on time in July 2012.

There were a number of stakeholders involved in the project and Mark Smith, project director at Turner & Townsend, believes that managing these disparate groups was the biggest challenge. Aside from the football community such as grounds staff, players, coaches, managers and club representatives, the project also included stakeholders from architects, the hotel provider, the medical operator and the national bodies representing sport. Electrical works also affected residents in a nearby village. 'All of these different parties are interested in their particular areas, but not the entire picture,' Mark said in an interview for *Project* magazine in October 2012. 'Managing all the stakeholders has been hard.'

The project team worked with different stakeholder groups to capture requirements both at the start of the project and on an ongoing basis as the project evolved. 'I'm not a football expert and I don't really know what's important to a footballer,' Mark added. 'It's very easy to dismiss something they think is important to them, when actually, they are critical. One of the lessons we all learnt is to listen to these people as they're the ones who are going to benefit from it, and if it doesn't work for them, we've failed.'[46]

During the scoping stage, your objective is to define what the project will encompass. It is not the place to get bogged down listing technical requirements and the colour of the wallpaper in the new office building. The scope of the project is what will be covered or touched by the work. Where does the project start and finish? Consider the following:

- Be clear. As with all project documentation, use the most specific words possible. If in doubt, clarify. Does 'launch' a new intranet mean just design it, buy the hardware and software, build the technical infrastructure and switch the site on? Or does it include producing training material, running courses on how to use it for every department and writing communication material?

- Document what is not included in the project. If you only focus on changing the employment contracts for staff in sales, say so. 'Updating the contracts for all other staff' should be listed in a section of the document titled 'out of scope'.

- Clarify that you have understood what your sponsor wanted to achieve by the project. Book a meeting with them and any other key stakeholders with the objective of verifying the scope statement. Send them the document in advance: they might have time to read it and reflect, coming to the meeting better prepared but plan to walk them through it anyway as the reality is many stakeholders won't make time to read papers sent out before a meeting.

There is plenty of research that shows that involving users has a direct relationship with how satisfied those users are at the end of the project.[47] Getting them involved is not always straightforward. The sponsor will often see it as your responsibility to define and document scope and will need convincing of the value of involving the end users at this stage. However, projects are done for (and sometimes to) the end users. Project managers have the luxury of moving on at the end of the project and not having to live with the consequences.

Having said that it is not a good idea to pull together a scope statement by yourself, it **is** a good way to get started. Call a meeting of all the relevant parties and present them a 'strawman': the draft scope statement. The objective of the meeting is to get commitment and input from those people who will be directly touched by the project. Ask the question: If we did this, what would be left out? What would we do that is really unnecessary? Be prepared to offer some suggestions to start them thinking. A good exercise is to pair up those present and ask each pair to try to break down a phrase in the scope statement into at least three more specific phrases. Then ask everyone to share and discuss and crucially, agree, their results. By the end of the meeting you should have a much clearer idea of what your stakeholders actually want from the project.

> Write your scope statement on flip-chart paper. You want it to have the air of a work in progress. Giving each attendee a copy of a professional-looking document, properly typed-up, will actually discourage people from suggesting changes: the scope statement already looks finished, and it takes a vocal team member to volunteer how it could be improved. This is one time where sloppy presentation is a plus!

AN EXERCISE FOR EXPLAINING THE IMPORTANCE OF FULLY DETAILING REQUIREMENTS

If your users see defining requirements as straightforward and the process of documenting them a waste of time, they are probably not thinking deeply enough about what the project entails. This simple exercise can be used at the beginning of a requirements gathering workshop to explain why it is important to write down each requirement in what can seem a painful level of detail.

1. Ask the group how difficult it would be to programme a robot to cross the road at a set of traffic lights and gauge their reaction.
2. Ask how long it would take to come up with the sequence of decisions and actions they would use to programme the robot.
3. They will probably suggest a time in minutes for this simple task. Give them as long as they have asked for to actually do it, but no longer than 20 minutes. Split them into pairs and ask them to write down a list of decisions and actions to get their robot to cross the road.
4. At the end of the time, ask each pair to report back and encourage the other pairs to critique their list.
5. Inevitably the robot will:

   - fall off the pavement because it forgot to step down;
   - get hit by a bus that didn't stop at the light;
   - walk over the person in front of it;
   - walk into an oncoming person;
   - get wet because it is raining and short circuit;
   - not know when it has reached the other side and trip over the kerb;
   - not be programmed to cope with snow;
   - and so on.

Do the exercise yourself beforehand and come up with as many conditions as possible that you think your user group might not consider.

This exercise is usually enough to get users to understand how complex requirements can (and should) actually be. You can then move on to discuss the project requirements themselves, encouraging the group to be as specific as possible to avoid a malfunctioning robot.[48]

Another reason to get users involved is that it minimises the mental model mismatch (see Figure 25.1). This term refers to the syndrome of users not getting what they want because what they described at the outset was transformed in the minds of those who actually then developed it – and by the time it got back to the end user it was subtly different from what was originally requested. Every person involved changes the end result just a little, as they have their own interpretation of what that end result looks like. The mental model mismatch happens for two reasons:

- The requirements are not explained well enough, so what the end user thinks they have described and what the person recording the requirements thinks they have described are slightly different.
- Those tasked with delivering to the requirements feel they can improve on the original ideas slightly, so they change things just a little as they go along.

Getting users involved in this early stage of the project will counter both these possible causes of the mental model mismatch. Firstly, they will have plenty of opportunities to describe exactly what they want and be part of the documentation process. Secondly, the project team who will end up building the new IT system, or whatever the project will deliver, will have a chance to meet users before it gets to the testing phase and really understand what they want. And to appreciate that just because a new tweak is possible it doesn't mean that it should automatically make it to the final deliverable unless the users agree that it really is an improvement.

Figure 25.1 Mental model mismatch

Include as much detail as possible, with input from your users and stakeholders, to ensure you do not leave anything out of the project scope statement.

26 COMMUNICATE AND DOCUMENT CHANGES

Despite having a clear and precise idea of what your project will deliver, agreed to by all parties at the beginning of the project, it is inevitable that there will be changes. When this happens, you need to be certain that you take the opportunity to explain the new changes to everyone concerned. Communicating **to** stakeholders needs a different set of skills to extracting information **from** them.

DOCUMENTING SCOPE

Liz Kirby managed a £6 million project to consolidate intranet sites from 81 countries into one consolidated e-commerce presence for a major telecoms company. 'We wanted to have a single content management system, a single search engine, one set of processes, and the organisation to support it all,' Liz explains. 'The project had implications for people's jobs – the intranet was going to be managed by a new UK team, so people in the territories were writing themselves out of a job.'

That wasn't the only problem that the project had: the scope was not written down. 'It was difficult to keep a grasp of the scope,' she says, 'as it kept growing. If the business was prepared to throw money at a problem, the scope just expanded to include the problem. We had written a terms of reference document and thought there was a common vision but everyone still had different interpretations of what was supposed to happen.'

This inconsistency in the scope affected the project communications. It was hard to make it clear to the end users what the 'launch' of the new intranet site actually meant. The inconsistency also affected the way in which IT and the intranet teams around the world interpreted their part in the project. Communications issues came to a head eight months into the project, when Liz realised that she had to act to ensure everyone had the same message. She sent members of her project team out to visit regional offices around the world. She herself went out to Japan and spent a few weeks with the intranet team there to clarify exactly what the project would deliver and what she needed them to do.

As well as validating the project progress, this approach had some other spin-off benefits. 'With a clear scope, you can trust people to go off and do what needs to be done without having to monitor the detail,' Liz says. As her project team was only 15 people it was essential that she could trust the local teams. 'It was also easier to

> resolve conflicts later on as we had built up relationships with the people involved around the world. And as they had not worked on anything like this before, it was a useful learning experience for the team and they can take that with them on to future projects,' she adds. 'It was difficult to manage the in-project communications over the two-and-a-half year project because of the scope changes,' Liz says, 'but harder to manage the messages to people outside the project. If the messages change, you just don't look credible and that doesn't help anyone.'

Communication during any project is important but when elements of the project are changing on an apparently random but regular basis it becomes critical to get the right message across in the right way. Joanna Goodman and Catherine Truss studied two major change initiatives, one at a company undergoing significant restructuring, the other at a company moving office location. Very few managers or employees felt that the communication around these projects was adequate. Goodman and Truss present a 'best approach' model developed from their academic research as well as the experiences at the two organisations they studied.[49] This model, the change communication wheel, is shown in Figure 26.1 and although it was designed to help plan communication strategies about change programmes as a whole, i.e. to a very wide audience, it is also relevant to in-project communications and the communications to your project team. The model illustrates the four elements of communication where a decision is required from the project manager:

- What is the message?
- How will we present it?
- How will people hear about it?
- What is our strategic approach?

The appropriate response to each of these questions depends on four external factors:

- Organisational context: your decisions about communications channels and messages will depend upon the situation within your organisation, as what would work in one company may be disastrous in another.
- Purpose of communication: the best approach will depend on what stage you are at in the project, as the aim of your communications will differ as the project progresses and the audience reacts to previous messages.
- Change project characteristics: different strategies and decisions are required for different types of projects. What would be suitable for the communication of a new bonus scheme would not be relevant for an office closure.
- Employee response: consider how you want employees to respond and also how you will find out how they have actually responded. This means building a feedback mechanism into your communication strategy.

All these elements must be taken into account when designing your approach (or approaches) to communication. Gauging the response to your communication is especially important when the message is that something within the project has changed, such as part of the scope. If this change has an impact on the work your project team needs to do, you must be confident that they fully understand the new status quo. Eddie Obeng, in his book *Perfect Projects*, writes:

> Imagine you say to someone 'Do you understand?' – what answer are you almost inevitably bound to get? If they understand, they will reply, 'Yes.' If they don't understand, but they don't know that they don't understand, they will still say, 'Yes.' If they don't understand, but are too embarrassed to say so, they will still say 'Yes!' In some cultures, the only acceptable answer anyway is 'Yes.' [...] So what question should you ask? One of a select group. Ask instead, 'What are the implications for you? What are you going to do as a result of what I've just said? How will this affect you next?'[50]

Using open questions like this gives you the confidence that they have understood and taken on board the message, even if they don't necessarily agree with it. The responses to open questions have the added advantage of offering you the opportunity to correct any misunderstanding at this point and not three weeks later when they deliver something completely different to what you were expecting.

> When sending out an email, use a descriptive subject line, not 'January Project Report' or 'Re: Your Question'. Even if the recipient does not have time to open the email and properly digest the contents they can at least get an impression of the situation from subjects like 'Project Alpha delayed by two weeks' or 'Project Beta passed audit'. For very short emails, put the whole content in the subject line, and end it with 'eom' for 'end of message'. The subject line of an email could be 'Bug 628 now fixed eom' or 'Are you dialling in to conference call? eom'. Instant messaging applications are better for this sort of short message, but you can use email software creatively like this if you don't have an instant messaging tool.

Communicating changes can also be a test of your credibility as a project manager. It can be difficult to maintain credibility when you have changes to scope or other elements within the life cycle of the project, as people can see you, as the representative of the project, as moving the goal posts. Trust is an important factor here, and when your project team is split over multiple locations, even within the same building, maintaining a level of trust within the team will allow you to maintain the project's credibility. It is far easier to write 'build trust within the team' than it is to explain how to start doing this as trust is something that develops over time. Honesty and predictability – doing what you said you would when you said you would and behaving in a way your colleagues would expect – are two factors which will provide a starting point for you to work out what trust means for your team.[51]

Avoid a one-size-fits-all approach to communicating change. Your project team will need a different level of detail to your sponsor. Equally, those affected by the change will need a different message to the one you give your boss. Not different factually, as inconsistent messages will damage your project's credibility, but presented differently and with a different amount of detail. Make the time to give the detailed version to anyone who asks, but tailor your communication to suit the needs of your audience.

Figure 26.1 The change communication wheel (adapted from Goodman and Truss, 2004)

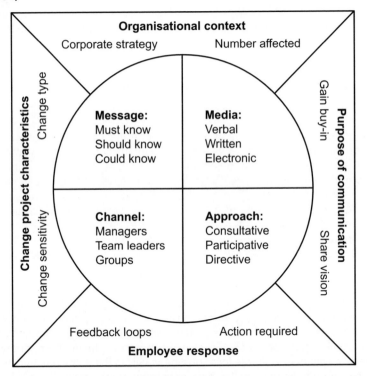

Projects change the status quo and projects themselves change as they progress. Effective communication is essential at all points through the project life cycle to ensure understanding and maintain credibility.

27 PLAN FOR HANDOVER INTO PRODUCTION

Project management is by nature a transitory job. You work on a project, it is completed and you move on to something else. Someone else picks up the burden of managing the deliverables on an ongoing basis – using the software, running the new office, or whatever it might be.

As tempting as it is to drop everything and walk away as soon as you've received sign off on the project closure document, your responsibilities don't end there. In fact, the handover to the production operational environment is also part of the project. Often this transfer to operations is done badly and not prepared early enough – if it is done at all.

DEALING WITH A POOR HANDOVER

When Phil Peplow was head of service delivery for a global financial institution he regularly dealt with incoming projects that were poorly handed over. One such project was a major change to the retail business software in use at the time. 'On implementation, certain features did not work, certain regional requirements had not been included, and testing had been curtailed due to looming deadlines,' he explains. 'When the project went live, the supplier's support operation was sadly lacking, the knowledge of the new system being within the heads of one or two individuals.' There was also no project handover documentation, leaving the operational team with a major problem.

'The project team had more or less been disbanded by this time and contract resource used by the implementing organisation had been let go,' Phil explains. To make matters worse, the project team had signed up to a support contract that only covered Monday to Friday between 9 a.m. and 5 p.m. If they had consulted the operational team responsible for delivering support to the end users, they would have realised that the software was in use for far more hours than that. 'Angry users took out their feelings on the only IT interface they recognised – the service desk,' Phil says.

The company had a history of poor project implementations which had left the operational team resigned to accepting the worst. 'Within the projects division there was a feeling that service delivery was the poor relation and was less competent than the project team,' Phil says. 'These undercurrents created fertile ground for a culture of isolationism. There was little appetite to engage with each other in

> order to craft any kind of joint working group until a "needs must" approach led to conversations and actions which were always too little too late. Projects must involve service delivery from the start in every case. Engagement is the key. No matter how little information is available, it can be shared and clarity will start to emerge through continual engagement. Clarity leads to understanding, understanding leads to confidence and confidence will help us to (jointly) deliver a good result.'

Projects have, after all, a start, a middle and an end. Being able to plan your exit is an important part of the end, and involves handing over your role as the driving force for the deliverables to someone else. As you will have spent a fair amount of time working on the project you will be widely considered as a (if not **the**) recognised expert, and even after the work is completed people will still direct their questions towards you. Unless you want the project to turn into your new day job, you will need to invest some time in planning how you will bow out gracefully and work with the operational team to complete an effective handover.

This starts with understanding that a successful project is only a success if things look good after implementation. In six months' time, will all the deliverables have fallen apart? For the end users, that's the definition of success. It's also what the operational team will want to know – are you giving them something fully formed and robust, or something that will be a recipe for disaster in just a few short months?

The best way to plan a handover is to plan for it from the very beginning of the project. Identify who will be responsible once the project team is disbanded and try to establish early on the kind of things they will want to see as part of the handover. This might include project documentation and access to the project wiki, for example, so you can start preparing these with the end, and future user group, in mind.

In the case of IT projects, the group who is likely to receive the handover, documentation, wiki and so on is often the service delivery team who run the business-as-usual operations. This group of people typically like things to stay the same. They provide stable, well-controlled and managed services to the rest of the business. Projects change that by introducing something new, so be mindful of the fact that you might hit resistance.

A project handover is traditionally something tacked on at the end of a project, as you can see from Figure 27.1.[52] However, a better way to manage the transition to production is to involve the operational team more and more as the project progresses, to the point where they are practically dictating how they wish the transition to work. If this happens, don't worry, even if it feels like this section of the project is now outside your control. They know what they need to make the project deliverables a long-term, stable success, so engage them in coming up with the most appropriate approach for the handover. The idea for both the project team and the production team is to eliminate surprises and to guarantee the success of the deliverables over the long term.

Figure 27.1 Increasing operational team involvement

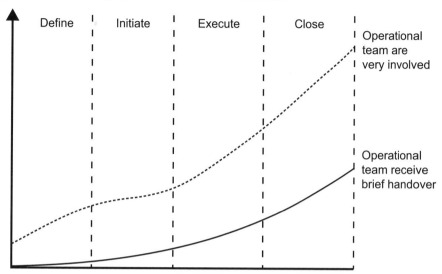

Make your exit an inherent part of your project planning, and tie this to the operational handover. When you know it is time to move on, go quickly and with a full handover to the business team responsible in the long term.

28 ACTIVELY MANAGE REQUIREMENTS

Requirements are the cornerstone of projects – after all, a project is all about delivering something new, and that tends to be documented in the form of project requirements. Unfortunately, poor requirements definition is something that is often cited as a reason for project failure.[53] Requirements need to be actively managed beyond capturing the initial requests.

Gina George is a project manager for a call recording division of a professional services firm in the United States.

'Ideally, in our organisation, the solution engineer's role is to do the discovery, document the requirements, and design the solution that meets those requirements,' Gina explains. 'My job begins when it's time to implement that solution. In reality, I would say that 25–30 per cent of my projects hit obstacles due to incomplete discovery and documentation of requirements. It is my job to fully document the requirements for the project as they were actually implemented.'

Gina records the requirements in a spreadsheet stored with her project documentation, but key information is also transferred to the helpdesk system for use as the company continues to support the customer after the project is completed.

As with many projects, the requirements for Gina's projects change as the project progresses. She manages the changes first by discussing the agreed system configuration and project scope as contained in the project documentation during her initial call with the customer. 'I ask them for confirmation that the hardware, software and services we've discussed match their understanding of the project's scope,' she explains. 'If there are adjustments, which are usually minor at that point, I reflect them in the working copy of the documentation that I deliver after the call. I also let them know that any subsequent changes will require a written change request and possibly additional cost.' About a third of Gina's projects hit unforeseen problems which result in changes to the requirements, managed through change requests.

In Gina's role, many of her projects are able to use a core standard list of requirements. However, she cautions against relying on the fact that one project has the same requirements as another. 'For my projects, I'd say 80 per cent of each

> project is similar, but the devil is in the 20 per cent that's different,' she says. 'The similarities do allow for some process streamlining, but major issues can arise when assumptions are made that something is similar when in fact, it's not. Case in point: I had a recent project where an assumption was made by the solution engineer that the VoIP recording required was a standard integration. As I uncovered during the initial project call, it was not, and our company had to make several monetary concessions in order to keep the project moving.'

Gina isn't alone in inheriting poorly defined requirements. Many project managers find that the main project requirements are documented in the business case or other pre-project paperwork and then handed on to them to deliver. Unfortunately, these are rarely adequate to create a fully fledged project plan.

Sit down with your project sponsor and other key stakeholders and ask lots of questions. What do they want to achieve through this project? Are they prepared to make the resources and budget available to deliver that? And if not, how will they prioritise requirements? Your other project team members also have a role to play in fleshing out poorly written requirements so get them involved too, especially when it comes to requirements in technical areas where you don't fully understand what is involved. Call on your subject matter experts to assist.

A starting point of documented requirements is exactly that – a starting point. Requirements change and evolve throughout the life of a project. Agile project approaches actively encourage change when this benefits the user and increases the usefulness of the end product. Even if you are not working within an Agile framework you can still keep on top of your evolving requirements by ensuring that any documentation is up to date and that the team know exactly what is expected of them when things change. You really want to avoid the situation of a team member continuing to develop to an old specification because he or she has not been told that the requirement has changed.

The flip side of this is that technical team members can **add** features to a project because they feel that the user doesn't know what is actually required or because they feel a small tweak will improve the functionality. Even if this is true, don't let your development team change the scope of what has been agreed with the project sponsor. Everyone should be able to put forward suggestions for changes, including project team members, so get them to raise their suggestion through the change management process and discuss it with the sponsor or project board before implementing it.

> Manage your requirements during the life of the project to ensure that you know exactly what is being delivered, even if things change along the way.

FURTHER READING FOR THIS SECTION

Bartlett, J. (2005) *Right First and Every Time*. Project Manager Today Publications, Hook.

Bing, L., Akintoye, A., Edwards, P. J. and Hardcastle, C. (2005) 'The allocation of risk in PPP/PFI construction projects in the UK'. *International Journal of Project Management*, 23, 25–35.

Cooper, D. F., Grey, S., Raymond, G. and Walker P. (2004) *Project Risk Management Guidelines: Managing Risk in Large Projects and Complex Procurements*. Wiley, San Francisco.

Hameri, A. (1997) 'Project management in a long-term and global one-of-a-kind project'. *International Journal of Project Management* 15 (3), 151–157.

Hancock, D. (2010) *Tame, Messy and Wicked Risk Leadership*. Gower, Farnham.

Harrin, E. and Peplow, P. (2012) *Customer-Centric Project Management*. Gower, Farnham.

Heldman, K. (2005) *Project Manager's Spotlight on Risk Management*. Sybex, San Francisco.

Pullan, P. and Murray-Webster, R. (2011) *A Short Guide to Facilitating Risk Management*. Gower, Farnham.

van Well-Stam, D., Lindenaar, F., van Kinderen, S. and can den Bunt, B. P. (2004) *Project Risk Management: An Essential Tool for Managing and Controlling Projects*. Kogan Page, London.

Williams, T. (2004) 'Identifying the hard lessons from projects – easily'. *International Journal of Project Management*, 22, 273–279.

SECTION 3: MANAGING PROJECT TEAMS

INTRODUCTION

> *The control of a large force is the same principle as the control of a few men: it is merely a question of dividing up their numbers.*
>
> Sun Tzu (6th century BC), *The Art of War*

Project management is about getting people with the right mix of skills together at the right time to deliver something in response to an identified need. Projects are done through people, so project managers need to be adept at working with others, especially as a project usually involves a group of people whose day jobs have very little in common but who have been thrown together to create a project team. As the manager of that team, you will probably not have line management authority over the people working with you, but you will still need to instruct, motivate, coach and cajole them into producing accurate estimates, progress reports and the project deliverables without taking too long or spending too much money.

Matrix management is not an easy skill, and handling the relationships within your team is what this section is all about. It covers managing your team, sponsor and project stakeholders, as well as other soft skills.

29 SET THE VISION

It is always easier to inspire your team about the lofty project objectives when the work is only just beginning (although obviously it depends on the project). However, you need to ensure their commitment for the long term, which means they do have to think about the future. The messages you offer at the outset will set the tone for the subsequent reinforcement of those messages throughout the project's life cycle.

Commitment and belief in the project's objectives by the wider team are seen by some project leaders as 'nice to have' extras, but you will find it a lot easier and more pleasant to work with a team who understand the project's vision and are prepared to work to achieve it. With that in mind, start as you mean to go on and aim to inspire the team from the outset and throughout the project.

WORK COLLABORATIVELY TO SET THE VISION

'I was recently made responsible for a large regulatory project in the financial services industry,' says Susanne Madsen, author of *The Project Management Coaching Workbook*, and herself a coach. The aim of the project was to build a centralised IT system which collates and sends information about over one million transactions a day to the Financial Services Authority (FSA).

Setting the vision for this project was a collaborative affair. 'I interviewed several individuals including the sponsor,' she says. 'I asked them what they ultimately wanted to achieve with the project, not only about how they wanted the new IT system to behave in the short term, but which user groups, product types, functions and geographical regions they wanted the system to cover five years down the line.' Susanne's conversations with her project stakeholders also included finding out about the benefits they wanted to achieve and how these could be measured. 'The project vision which we arrived at was to build a fully configurable system which could handle any type of financial product and any type of regulatory reporting requirements from any regulator across the globe. The vision is furthermore that the system will handle up to five million transactions a day and that rigorous data controls will help the company to save money on execution costs and manual interventions.'

Susanne has also worked on projects where the stakeholders were not so aligned about the project vision. 'I worked on a large financial project where the project sponsor and the primary user disagreed to what the vision was,' she says. 'The sponsor wanted the project to primarily generate cost savings for the department whereas the user representatives wanted the project to generate better working practices and controls. In spite of several workshops these two stakeholder groups agreed to disagree and never fully reconciled their views. In the end the project sponsor won the battle simply because she was in control of the budget. Project funding was cut and the project was closed down as soon as the desired cost savings had been realised.'

Getting key decision makers to agree on common vision is one thing, but getting the project team and the wider stakeholder community to believe in it is something different. 'I like to incorporate the project vision into the key project artefacts right from the start of the project,' Susanne explains. 'That means that I will describe the vision in the Project Charter or Project Initiation Document. On my latest project I also created a presentation with the sole purpose of describing the project vision, purpose and approach so that we could better communicate the vision to the wider stakeholder group and gain their buy-in. Every couple of months when we gave the presentation to someone new we would check with the sponsor to ensure that the vision was still accurate and update all relevant artefacts accordingly.'

Susanne believes that the best way to get the project team to believe in the project vision is to involve them in the project and expose them to the key stakeholders from day one. 'When they are involved they feel that they have a stake in the project and they will be more driven and motivated to achieving and contributing to the vision,' she explains. 'For this reason I often invite team members to attend stakeholder workshops and I also encourage direct contact between a team member and an end user or stakeholder when the circumstances are right.'

In the early days of your project you will probably find your team hugely committed to achieving the objectives. But there's a risk this will wane as soon as their day jobs start to take precedence over project tasks. Realisation that the work cannot be completed instantaneously will quickly set in. As the project manager, you have the responsibility of keeping your team focused on and dedicated to the job in hand. This is a task made more challenging by the fact that some of your team may not have chosen to work on the project willingly and are just doing what they have been told. Making them believe in the long-term success of the project is the secret to ensuring a good working relationship with your team, and in achieving your end goal. Do this through a clear project vision.

It is not always easy to create a clear vision for the future, especially if your project sponsor refuses to think about the long-term goals. A project may have very different objectives for the coming year and the coming five years. Keep digging and take every opportunity to press the senior stakeholders on what their long-term vision is for this project so that you can incorporate this into the project communications.

Take time to explain the goals of the project: do not assume that because their boss has allocated them to the work that he or she has also explained the ultimate objectives. You know why the project is important, so translate that into reasons why it is important for the people working on it. Put the project into the context of the company's overall strategy so individuals can see the link to the bigger picture. Will it make their daily lives easier? Help make training new staff simpler? Will it reduce customer complaints? Or generate more sales?

One discussion at the start of a project may not be enough to keep the commitment level high, especially when the project hits difficulties or is perceived to be taking a long time – or if key stakeholders don't agree on the vision and you are dealing with conflicting views. This can negatively impact on the project team's morale and the progress of the project, so try to tackle this conflict by asking questions and facilitating a discussion until you reach agreement on one, solid, vision for the project.

Enlist the help of your project sponsor when it comes to setting and reinforcing the vision. If the main stakeholder or project sponsor attends a meeting, invite them to comment on how the project work will help achieve the vision. Support from senior managers is a good way to practically demonstrate corporate commitment to the project. The team will be more committed to the project themselves if they feel that their work is being recognised and supported from those higher up. Get the sponsor to drop an email to the team thanking them for their achievements so far and reinforcing the end objectives. Better still, ask the sponsor to mention your project in his next company briefing so the message that your project is important reaches a wider audience. Support from senior management is the second most critical factor in project success, after having clear goals and objectives.[54] If you can't reach into the upper echelons of the company, ask business users who are already starting to see benefits to attend a project meeting and explain the difference the project was made to them.

Equally, you can keep reminding your stakeholders and project sponsor about the vision by linking the project plan and deliverables back to the vision whenever you share project information with them.

> Try to create a culture of commitment and belief in the project's vision at the outset at all levels, and continually reinforce this throughout the life cycle of the project.

30 KNOW THE CULTURE OF YOUR TEAM

Every team has its own way of working and when you start working with new people you need to appreciate the existing culture. Each team member will already have preconceived ideas about you and the others, at an individual and departmental level, before you even hold your first meeting. Trying to work out and understand these views will help you appreciate the culture of your new team and how best to make them work together successfully.

WORKING WITH MULTI-CULTURAL TEAMS

Adrienn Unachukwu-Hamori has worked in a project environment for 15 years and is a European project manager for a multinational company. Originally from Hungary, she now lives and works in the UK, so she has plenty of first-hand experience of multi-cultural teams.

'There are common features with the projects I have worked on, with teams made up of people from different countries and sometimes different continents. But twins aren't the same and neither are people from different cultures,' she says.

Adrienn says that the most obvious difference on a multi-cultural team is language. After that, she believes the next biggest difference is in business etiquette. 'In some cultures people just bow, in Europe we mainly shake hands and in some places a hug is also acceptable.'

Then there are differences in response to hierarchical structure and humour, religious differences and culinary differences at business meals. Time management is also a common difference between cultures. 'I wasted so much time waiting for people,' Adrienn says. 'It is very difficult to make clear that everybody's time is expensive and we have to keep to the timetable. The time zone difference also causes problems. Many people don't successfully manage the different time zones – I've received many phone calls during the night.'

Adrienn believes that it's important to set the right tone at the beginning of the project. She tries to schedule an informal dinner along with the kick-off meeting so that the team members can get to know each other.

She has started to add a new activity to the kick-off meeting as well. 'I ask the team members to tell everyone their expectations and fears about the project,' she explains. 'It helps not only me, but also the other team members to co-operate.'

During the kick-off meeting Adrienn clarifies what she calls the project rules. 'This clarification gives everybody the opportunity to discuss special requirements like daily routines, reporting requirements, meeting schedules and team responsibilities.' It is also an opportunity to learn about the cultures in the team. 'This gives me guidance on where I have to be more focused and patient. I had to learn not to surprise people with things, but accept that everybody is different and what's obvious for me could be totally strange for others.'

Adrienn also uses tools to make life easier. 'I use a world clock to know exactly what the time is in different countries,' she says. 'Our company uses an intranet, so I manage and record the project there. This way everybody can get the necessary information anytime, and they can update the project anytime. It saves a lot of time and extra work.'

Finally, she warns that from her own experience ignoring cultural differences can make everything on the project harder. 'Cultural differences cause delays, costs and contribute to a stressful life. You can save time, energy, money, and lead the project more effectively if you pay attention to and accept the differences of people,' she says. 'If you, as a project manager, don't manage the challenges of team members from different cultures, every moment of the project could be a problem. The team members won't be free and they won't be able to work effectively. They won't know how to behave themselves and the environment will be uncomfortable. It's your job to manage that.'

Fons Trompenaars and Peter Woolliams define culture as:

'A series of rules and methods which a society or organisation has evolved to deal with the regular problems that face it [...] Culture is to the organisation what personality is to the individual – a hidden yet unifying theme that provides meaning, direction and mobilisation that can exert a decisive influence on the overall ability of the organisation to deal with the challenges it faces.'[55]

It is precisely because culture is so ingrained in the way people work, and the way a company is organised, that it is sometimes difficult to spot what a team take for granted about how they do their jobs. Even a project team made up of people of just one nationality will have a particular culture, evolved from the organisation's own corporate culture.

Corporate culture manifests itself in many ways. Do team members gel quickly and spend every Friday evening at the pub? If appropriate, consider going to the pub a few times so they get to know you. Do they keep information to themselves? Foster a culture of sharing information by encouraging them to outline their contributions to the project at your team meetings, and help them to see the advantages. Does the company have a culture of long tenure? Do they come from departments with strong hierarchical structures? If they have held their positions for a long time be aware that they might resent you setting them tasks. Get used to making a quick call to their line managers so you can approach each new activity with, 'I've run this phase past X and Y and they agree the next steps are....' And even if you can identify what makes your newly formed team tick, how will this help you manage the project?[56] Being aware of the preferred style of your team will help you relate to them, and will help you understand how to get the best from them.

> Most project managers work with teams over two or three locations, with less than a fifth managing a team all based in a single location. Distributed teams come with their own challenges, such as managing different time zones and introducing technology to keep everyone aware of the project's progress. Only five per cent of people report that working with distributed teams is easier than co-located teams, with 80 per cent saying that managing teams over multiple locations is harder.[57]

> Cultural awareness, whether cross-country or within a company, will help you get where you are going more smoothly.

31 AGREE WHO IS GOING TO SIGN OFF

Who on your project will have the authority to make decisions at the most senior level? If you do not yet know, find out before the project gets too far downstream. You need to have someone to turn to in case you need a decision on the way the project is heading, and political power games can significantly delay (or stall or close down) a project completely. Thinking through the organisational structure of the project in advance can help prevent that.

TAKING THE LEAD FOR VETERANS

In 1998 the US Department of Defense (DOD) and the Department of Veteran Affairs (VA) started to work together on a project to share medical data for active-duty and veteran military personnel. The rationale behind this work was that as service personnel were highly mobile they ended up with medical records at military facilities all around the world. An electronic system to share information between the two departments would help service personnel get the best care, and also help process any disability claims.

The Government Computer-based Patient Record Project was started, but by April 2001[58] it was struggling. The US Government Accountability Office (GAO) carried out a review and concluded that the departments involved needed to agree who was going to take the lead and be the main decision-making authority. This would establish a clear line of authority for the project, which would in turn make it easier to draw up and follow comprehensive plans.

By June 2002 the departments had still not agreed who was in charge, and the project was moving forward without co-ordinated plans, performance measures or clear goals and objectives.[59] The GAO once again recommended that a lead agency was designated to give the project a route for decision making.

Two years later the GAO reported again that it was hard to establish what progress had been made with the two-way data sharing project as no lead entity had been identified, meaning that no department had the right to make decisions that were binding on the other.[60] The initial aim of making military medical records easily accessible had so far taken six years, and still seemed no closer to getting agreement or leadership with the way forward.

> After reading the draft 2004 GAO report, the Secretary of Veterans Affairs documented the department's response to the recommendations. He stated that the VA/DOD Health Executive Council would be the body that provided the final decision-making authority. However, by May 2005[61] no agreement had been reached regarding data sharing, leaving the VA uncertain that vocational rehabilitation services were being provided to all service members who needed them, and both departments were still struggling to implement an IT system seven years after they both agreed in principle that it was a good idea.

> It is essential to work out who on your project will have the authority to make decisions at the most senior level. You need to know who will be able to make the final call when you are asking for a decision on this solution or that, increases in budget, taking staff out of their day jobs for training and so on. Don't wait for an important issue to arise before you attempt to work out who in the organisational structure has the authority you require.

Break the questions down and you might find you need it to be several people: your sponsor may agree that staff training is necessary, but it could be the head of the operational department who decides when to make the relevant staff available. A survey by the Centre for Complexity and Change at The Open University found that a third of project managers are involved in the decision-making process but are not the sole decision maker themselves. It also found that 24 per cent of respondents had no influence over the decision-making process at all.[62] If you fall into that category it really is critical to work out who you need to turn to for decisions, so that you can start to put together a stakeholder management plan and try to gain some input into the decision-making process, at least in terms of your recommendations being considered seriously.

The decision makers on your project should be authoritative and have the ability to negotiate with other key departments in case of conflict. Without a clear route for (binding) decisions, you could find your project stalling at the most critical moments, as the stakeholders disagree among themselves and refuse to abide by each others' decisions.

> Work out who will be able to make the binding decisions on the project in advance of needing to ask that person to do so.

32 DON'T FORGET THE SOFT STUFF

Project management is largely about hard skills: dates, money, resources, ticking off tasks. But to do those successfully, you need an appreciation of 'soft' skills: those elements of project management it is too easy to overlook when faced with time pressures and other crises. Soft skills are also important in a project environment because frequently the project manager will not have line management authority for the people working on the project.

GATHERING FEEDBACK – AND ACTING ON IT

At one UK healthcare firm, the IT project management team realised that they could probably do better on their projects – if only they knew what their project customers wanted them to improve. They realised that waiting until the end of the project to carry out a traditional post-project review meant that the feedback that was provided came too late to proactively do anything about it. While it was useful to know for the next project, there was rarely a high degree of overlap between projects so the lessons learnt were archived for posterity rather than for action.

The team implemented a simple survey that they could use monthly to gather customer feedback throughout the life cycle of a project, not just at the end. Project stakeholders were regularly asked for their comments on four measures:

- How well the project team managed issues.
- How well the project team communicated.
- How well the project deliverables were being integrated into the existing hospital processes.
- How well the project was going overall.

For each measure, the customer was asked to score the project out of 10. They were also asked to identify their top three issues.

Project managers could then take this regular feedback and act on it, resolving the top three issues and aiming to improve the scores each month so that they (and their stakeholders) could see regular improvements in the way the project was being run. This approach enabled the project managers to improve stakeholder engagement and demonstrate that they were actively listening and acting on any concerns.

> This regular, constructive approach to feedback enabled one project team to move a stakeholder's score from 5 to 9 out of 10 over the course of 10 months through regularly reviewing expectations and improving process.[63]

'Soft' skills are as much a part of the project manager's job as making sure tasks are delivered on time, and as much a part of the overall success as your end product. They include:

- stakeholder identification and management;
- understanding the culture of the team;
- putting together and using a communications plan;
- being present and available for your team;
- giving them praise and guidance;
- getting buy-in from your team and also from the senior management – in practice, not just on paper;
- understanding who is accountable for decision making;
- ensuring people know why the project is happening;
- offering training to back up the communications if necessary;
- being positive and an agent for change yourself – never let anyone catch you saying the project is a waste of time!

Flick through the rest of this book for examples of these skills in practice.

Improving your soft skills will help you get the best out of your project team. Academics have long been searching for the magic bullet that makes one project team more successful than another and even within a single organisation with a common culture and well-motivated staff some teams will just perform better than others. Marla Hacker has researched the theory that the individual characteristics of team members and the dynamics within the team both influence how the team will perform. She studied 22 teams made up of three or four people drawn from a group of university engineering students. The teams worked on a project over a semester and were graded in relation to the other teams, providing an incentive to perform well. To analyse whether individual characteristics made a difference to the team's overall performance, Hacker asked each student to complete a questionnaire about themselves and their experience. For the team dynamics part, the students rated their team's performance on 12 factors including quality of discussion and level of agreement. Hacker's research concluded that only one factor made a difference to team performance: the students' average academic results, which she believes is representative of ability to accomplish tasks combined with the students' ability to motivate themselves.[64] She found no link between the students' demographic background, experience or team dynamics and team performance.

Project managers often wish they could handpick their team to get the 'best' performers. Hacker's research shows that it does not matter whether you choose your team or have them allocated to you because their individual social and employment backgrounds, and how well your group dynamics work, will make no difference to the team's ability to perform well. While her study was not carried out in a workplace environment, an appreciation of soft skills and team interactions will certainly make your job easier.[65]

> Balance the technical aspects of project management with paying attention to the 'softer' elements to help embed the change and support the project team with the implementation.

33 TRAIN YOUR SPONSOR

You might be lucky enough to end up working with a sponsor who has already championed many varied projects, understands their role exactly and is keen to support this project and the team. Or you might find yourself working with someone who doesn't really understand what is required of them. If you find yourself in this situation then you need to take your sponsor under your wing and help them understand how their role supports the project team.

CUTTING THROUGH THE JARGON

In the 1990s Neville Turbit began working with a major Australian government department to implement a project management methodology and tailor it over time to meet their needs. 'It was a big organisation where IT and the business did not enjoy the most harmonious of relationships,' he says. The department suffered from the same problems of many other large organisations at the time: prepare a project specification, and throw it over the fence for IT to build. 'IT built it and, surprise, surprise! It was not what was needed at the time of implementation,' Neville adds wryly.

He knew that implementing a methodology alone was not going to solve the problem. 'I adopted a two-pronged attack on the business,' he says. 'Firstly, we had to train sponsors to be involved, and see themselves as decision makers. Secondly, we had to convince the rank and file business users that unless they were involved all the way through the process, they would never get a satisfactory outcome.'

Neville, the principal of Project Perfect, a project management software, consulting and training organisation based in Sydney, Australia, started with the management population who would soon become project sponsors. His 15 years of IT consultancy and a similar length of time in business roles has given him an insight into how sponsors react in project environments. 'In my experience,' he says, 'the biggest problem is getting the person who is responsible for the project to understand what their role is all about.' Neville feels that sponsors are out of their normal operating environment when they are working on a project, and that can block them from being effective. 'Many sponsors have little idea why a project is different to any other department they might be responsible for. I have had one sponsor say to me: "I don't know what you want from me. You are more demanding than any of my line managers." I had to point out that I was doing a non-routine job. His line managers had clearly defined processes and procedures and they just tweaked the business process. I was creating it.'

One sponsor confessed to Neville that she had no idea of what she should be doing. 'She felt so uncomfortable she basically delegated all the decisions to IT,' Neville says. 'When I first talked to her about using a methodology, it was not much better. Once again it was a language that didn't sink in.' Neville continued to work with the sponsor, helping her understand her role and the 'world of projects'. 'The breakthrough came when I took her a decision on a project,' he recalls. 'I told her that she needed to make the decision. I said: "It is your responsibility. Our responsibility is to present you with options and not leave here until we explain the problem and the options to resolve it in language you can understand." Evidently we spent considerable time discussing the issue in layman's terms.' Neville stopped talking in project management jargon, gave up using the words 'databases' and 'functionality' and started talking about what the new system would do, and what things the team needed to keep track of. 'She said this was the first time she did not feel foolish in front of her subordinates discussing the new IT system,' Neville says.

Neville made sure the roll-out of the project management methodology at the government department included time for sponsor training. The focus was on getting project managers to talk the language of the business 'rather than techno-babble'. He also made sure project managers knew never to take a problem to a sponsor without a number of solutions. 'Over time we built up a core of people who sponsored projects. They knew the questions to ask, and were not afraid to become involved in providing solutions,' he says. 'It's taken a decade but this particular department now have the most involved business resources in any organisation I have seen. Project scopes are developed in an orderly manner, and budgets are realistic. It is always interesting when new people come into the organisation. Typically they have a better way of running projects. The first person to tell them to "do it our way" is typically a business person. The business people have seen the approach work, and the sponsors have sponsored projects that they truly controlled.'

Eddie Obeng in his book *Perfect Projects* defines the sponsor as a person who:

- invented the idea and really wants to do it
- controls the money
- wants the end product or will end up living with it
- can provide effective high-level representation, and smooth out the political battles before you get to them
- 'owns' the resources
- acts as an effective sounding board/mentor.[66]

Unfortunately, you might not be working with someone who meets all these criteria, but it is a myth that you need one sponsor for the duration of the project. On large projects it can be beneficial to have the sponsorship role pass between different senior executives. After all, the person best placed to assist during the planning phase with the strategic viewpoint and the ability to secure resources may not be the best person to support you during the execution phase when you need someone more hands-on who can focus on assisting the project team get on with their jobs.

People in the role of sponsor also suffer from having many different demands on their time and your project may not be top of their list. If they don't know what it is they have to do, you can be sure they won't be able to make the time to find out. You have to be there to help them discover what being a sponsor means and to explain what you expect from them.

Some organisations offer sponsor training and expect all their senior executives to attend. Where formal training is not available, or if it has been a while since your sponsor attended something like that, it falls to you to offer some guidance. The beginning of the project is normally the best time to introduce a sponsor to their role. Try to find out what experience they have had, what went well and what they found difficult about sponsoring previous projects, but be warned that senior managers may not be willing to share their experiences with you. You can still approach the subject tactfully: 'I know you've already sponsored loads of projects but as we haven't worked together before I just wanted to explain to you how I see your role as the project sponsor, and then we can establish how best we can work together.' Use the sponsor FAQ below or make up your own as a starting point for discussion.

SPONSOR FAQ

- Why does a project need a sponsor?
  To support the project team and act as an escalation route for any issues or problems.
- How is a sponsor different from a project manager?
  The project manager manages the operational, day-to-day issues on the project. When something happens that they can't manage within the agreed parameters (a budget, a time frame, a set of requirements) a sponsor makes the decision about how to proceed.
- How can they make the decision when they don't work daily with the detail?
  The project manager will present various options and the consequences of following each option. The sponsor should have a general overview of the project which will be enough to choose the right course of action.
- How is a sponsor informed of progress?
  This is agreed between the project manager and the sponsor at the beginning of the project. It could be a written monthly report, a face-to-face briefing or on an exception basis. If there is a problem that cannot wait, the project manager should be able to approach the sponsor immediately.
- What else does a sponsor do?

  - represent the project at a senior management level;
  - keep the project manager informed of any changes or developments that may have an impact on the project;
  - put their name to and help with communications about the project;

- offer advice and make decisions;
- put forward and/or support the case for a comprehensive budget for resources;
- chair the steering group;
- read, understand and approve project documents;
- anything else (within reason!) to support the project at the request of the project manager.

A sponsor may also need 'training' in the more technical elements of the project. As the work continues you will soon become an expert in the intricacies of what it is your project is delivering, but you cannot expect them to understand the details or the jargon. Present your project updates with clarity, keep the use of jargon to a minimum, ask open questions to test their understanding and give them the opportunity to ask you questions too.

Don't automatically assume your sponsor knows how to carry out their role effectively. Find out and explain what you expect from them.

34 CARRY OUT STAKEHOLDER ANALYSIS

One of the most important tasks of a project manager is to manage the stakeholders' expectations and to communicate with transparency about project status, risks, issues and decisions. Your project will touch the working lives of many staff. These are your stakeholders. You will quickly identify some of the groups affected, but there are probably others who have an equal influence over the success of the project, and it could take a while to uncover all these groups. Stakeholder analysis is a tool that helps identify all the relevant parties and map their interest and influence over the project.

INVESTIGATE THROUGH INTERVIEWS

When Susanne Madsen, author of *The Project Management Coaching Workbook*, took over a $30 million, three-year IT project she didn't know the company or the industry particularly well. She therefore had to spend a significant amount of time identifying the relevant stakeholders.

'The way I went about identifying all the stakeholders was to put on my "investigative hat" and interview anyone who I knew was connected to the project,' Susanne explains. 'In the first instance I asked my own manager, who was head of IT, to provide me with as much information as he possibly could about the project, the end users, the sponsor, the technology teams and all the decision makers he could think of – internal and external to the company.' This gave Susanne an initial overview of the main stakeholders.

'I then engaged the project sponsor and asked her a set of similar questions about the project and who the users and other interested parties were in addition to herself,' Susanne adds. 'When I subsequently spoke to each stakeholder, more stakeholders came to light which I added to my list.

Once Susanne had identified all stakeholders, she analysed each one in turn to establish their level of power and influence. 'I asked the project sponsor for information about each stakeholder and who the main decision makers were according to her,' she says. 'I also spoke to each stakeholder personally and used my sense of judgement to understand how driven they each were, how determined they were to make their voices heard and what their position was inside the organisation and vis-à-vis the project sponsor. From my analysis I also found out what each stakeholder wanted to get out of the project and their individual communication styles and preferences.'

While this analysis was useful, it wasn't until the project started in earnest and Susanne's project team needed to make important decisions that people's real authority came to light. It was through the weekly working group meetings and the monthly steering committee meetings that she understood who the actual decision makers were. These were the people who were good at challenging the status quo, making their voices heard and who were exceptionally good at persuading others – including the sponsor – of their views.

'I would not capture information about the level of power and influence of each stakeholder and my strategy for keeping them satisfied in a document, due to its sensitivity,' Susanne explains. 'I normally carry out this analysis on a piece of paper in private. After completing the analysis I would destroy the paper so that it does not end up in the wrong hands.'

Susanne says that stakeholder analysis helps her understand everyone's motives and it makes it easier to position herself and the project correctly in relation to each stakeholder. 'The analysis helps me to "decode" messages from stakeholders and to communicate in the most effective manner with them,' she says. 'I will often share my insights with other team members verbally so that they can also fine-tune the way in which they communicate with certain stakeholders. My priority will always be to spend my time with those stakeholders who have the most power and influence over the project.'

Stakeholder analysis or stakeholder mapping is usually done in three steps:

1. identify all the stakeholders;
2. record their position (interest) regarding the project;
3. compile an action plan for how to best influence them.

Identifying stakeholder groups is the easiest part of the analysis.

Stakeholders are 'individuals, groups or institutions with an interest in the project, and who can affect the outcome'.[67]

It will be straightforward to identify the inner circle of stakeholders: you, the team, the sponsor, the project's customers. You do need to go beyond that though, so draw on your team and use their ideas as well to make sure that all relevant parties make it on to the list. For example, stakeholders can also be external groups like government bodies, regulatory agencies or people from other commercial organisations. Capture the department or organisation they represent, their official role in the company and their project responsibilities and any previous experience that is relevant to this project.

You can also link stakeholders to the project success criteria that apply specifically to them or the groups they represent, and this can help with the next step – plotting their current interest in the project using a chart like Table 34.1. Don't ask stakeholders directly where they see themselves: they will probably give you a politically correct answer. Base your evaluation on what you see them do and say. Then work out where you want each individual or group to be in terms of their support for the project. You will never need all your stakeholders to be 100 per cent supportive all of the time. If your chart reflects that, you have probably missed out some of the smaller stakeholder groups.

At this stage consider the relative power and influence of each stakeholder. If all your critical, high-power stakeholders are currently not showing much interest, you have already identified a risky situation which needs addressing quickly. What can you do to address it? Communication, attend meetings, explain the benefits, find other people who can sell the project on your behalf, find allies, network and use whatever influencing techniques work for you.

To make the most of your stakeholders, move to step three: compiling an action plan to identify what as a team you can do to move each stakeholder from their current 'interest' position to where you need them to be. Table 34.1 shows that the seven sales support staff have the furthest to go in terms of changing their attitudes, but they are a group of low-power stakeholders. In this situation, the project manager would need to tackle this group, but would have to balance the effort required with ensuring the IT Manager, a high-power stakeholder, is also influenced, as this role appears more critical to the project's success.

You can include a lack of interest from stakeholders as a risk on your log if you want to manage your action plans formally. Track the actions required to influence each stakeholder as you would do for any risk management plan. Just be aware that the risk log is not a private document and it may be more astute to keep some of those activities just within the immediate team. Stakeholder analysis and the subsequent activities can be political and organisationally sensitive.

Your stakeholder analysis can also help your communications planning. Map the communication needs of each stakeholder to the appropriate activities on the communications plan. You won't forget to communicate to any stakeholders if you use your stakeholder analysis like this! It can also help you adapt your approach and style effectively to the needs of each stakeholder so that you can work towards a position where all the stakeholders support the objectives of the project and will end up considering it a success.

There is one further use for your stakeholder analysis. It will help you decide who should be on the project's steering group or project board. The sponsor will also have a part to play in deciding on the make-up of this group, but you should aim to ensure that all the important, decision-making stakeholders are represented in the project governance model.

CARRY OUT STAKEHOLDER ANALYSIS

One of the most important tasks for a project manager is to manage stakeholders' expectations and to provide transparency about project progress, risks, issues and decisions. Mapping your stakeholders' positions with regard to interest and influence will:

- quickly highlight any immediate concerns, and;
- allow you to draw up an action plan to address those concerns.

SIX STEPS FOR ONGOING STAKEHOLDER MANAGEMENT

1 Identify stakeholder groups

You can't start managing stakeholders until you know who they are. Think of the main groups or departments affected by your change project. Launching a product will of course touch marketing and sales but what about training? Opening an office will affect HR and IT but what about the switchboard operators and the internal mailroom? Carry out stakeholder analysis to understand their interest and influence over the project.

2 Nominate a key contact

From each of those groups pick someone to be the key individual. Be careful who you choose. What are the internal politics of picking one over another? Your key contacts should ideally be people who are directly affected, with enough authority to make decisions about things that touch their departments.

3 Feel what's going on

Contact your list of named individuals and introduce them to the project properly, dispelling any myths they might have held. Get an accurate understanding of their position regarding the project; validate the assumptions you made about their interest and influence in step one.

4 Observe them

The more you work with and involve your stakeholders the more chance you have to observe them. Don't let stakeholder groups drift away. As the project progresses, act to bring them further into the project in a positive way, observing how their attitudes change and adapting your action plan accordingly. Your aim is to work out each negative touch point and address it, moving the stakeholder's negative attitude to a positive one.

5 Review what's going on

People and job roles change. So do projects. The person who was suitable to represent the legal team 10 months ago may not be the right person today. Don't be afraid to ask your stakeholder if they still feel relevant to the project – and if they are not, ask who should take over from them. Brief the new addition to your project team on their role and responsibilities, decisions in the pipeline and what decisions you will expect of them in future.

6 Manage the process

Finally, monitor and manage your stakeholders and their expectations as the project progresses – not just at the outset and when you need something from them. A quick call every now and then (put it in your diary) just to keep them up-to-date will go a long way to promoting goodwill and building your reputation as an excellent project manager. At the end of the project, thank them and manage them out of the team. Managing someone out of the team means they understand that the project organisation is coming to a close. Each ex-team member or stakeholder should understand the process for raising future issues or questions through their new business-as-usual contacts.

The key message

The acronym formed by these six steps is INFORM. Ask your stakeholders at the outset how they would like to stay informed and make sure you do it. Do it even if they say they are not bothered about receiving information about the project. One day they will be interested and then they will know exactly how to reach you for more information.

Table 34.1 Mapping stakeholder interest and influence

Stakeholder	Averse	Unco-operative	Indifferent	Supportive	Enthusiastic
Sponsor				□————▶◆	
Finance director	□------------------▶◆				
IT manager		□ —————————▶◆			
Database analyst			□◆		
Training team (3 people)			□············▶◆		
HR supervisor		□--------------------------------▶◆			
Sales support staff (7 people)		··▶◆			

□	Current position
◆	Desired position
——▶	Highly influential stakeholder
----▶	Influential stakeholder
······▶	Uninfluential stakeholder

35 PRESENT YOUR STUFF INTERESTINGLY

As a project manager you will have to explain the details of your project to non-project managers, non-project team members and to people who have very little time to listen to what you have to say. By thinking about what you say and how you say it, whether it is for a short update meeting or an hour-long presentation, you can engage your audience more effectively.

MAKING YOUR MARK

Jo Ann Sweeney specialises in project stakeholder engagement and is the founder of Sweeney Communications. She worked on the SPARKS programme,[68] a lobbying and influencing campaign to change UK and EU driving and parking laws.

'We based choices about channels and materials on the needs of our different audiences – national, European and trade press, parking teams in local authorities across the EU, UK and EU politicians and the project team,' Jo Ann explains. 'At the heart of our choices was the objective we had for each piece of communication activity – did we want to increase knowledge and understanding, change attitudes and perceptions, or have our audiences commit to a specific action?'

When the objective was knowledge and understanding the team chose printed and online channels such as an electronic newsletter, website, emails, leaflets, brochures, media packs and streaming video. When the objectives were changing emotions or calls to action, they opted for face-to-face channels such as presentations, receptions, seminars, workshops and meetings.

'Communication activities that are focused on audiences – what they are interested in and how they like to receive information – don't usually bore the people they are targeted at,' Jo Ann says. 'For instance, our two media packs gave journalists and editors key background they've used repeatedly in news stories and feature articles since. Information was presented in succinct bullet form with graphs and images they could reproduce.'

Jo Ann, who typically acts as the communications lead on complex HR and technology projects for major corporations, says that events were handled in a similar way. 'When we invited audiences to events we clearly explained the purpose in the invitation and the benefits for them of attending,' she says. 'Some of our events were social with a small element of business as this is what worked best for

> some audiences; others were discussion and decision making events with a smaller networking element.'
>
> The project team had many discussions before disseminating messages, thinking through how different audiences might react and the best way to get the reaction they wanted. They also stayed in touch with the different audiences so that they could see how their knowledge, attitudes and likelihood of supporting SPARKS were changing. 'We would phone our key stakeholders on a regular basis to get their feedback, or go and meet them in their offices to discuss their issues and perceptions of the programme,' Jo Ann explains. 'Then we tailored messages and activities based on what we heard.'
>
> Jo Ann has three simple suggestions for managing project communications effectively. 'Tailor your messages and activities to your audiences,' she says. 'To do this successfully you will need to know quite a bit about them. Second, clarify anything that is complex so audiences understand the first time they listen or read, without dumbing down the content. Finally, disseminate information regularly and often using different communication activities so people stay interested.'

Whether you are presenting information to your team or senior management, bear in mind your audience and try to make your work interesting and memorable. However, be careful to avoid style over substance – in other words, your presentation should include all the key facts and prioritise these over showmanship. No one really cares if you know how to use slide builds, sounds and animations and these can overshadow your message.

You may not have the opportunity to change the format of standardised reports like monthly progress updates as changing the format of those could cause problems for the people who receive and use the information. You are, though, at liberty to be creative with ad hoc reports and presentations. Here are some ideas.

TIPS FOR PRESENTATIONS

- Make it professional, using your company or team's standard presentation layout. Not got one? Set up a template. Get someone to proof-read it: your audience should be captivated by your fascinating presentation, not on the lookout for the next spelling mistake.
- If you have to send out your presentation in advance, and then intend to talk through it at a meeting, be sure to have some extra information to give. Reading out what your audience has already seen is certain to create a few droopy eyelids.
- Avoid clip art-type pictures. These look unprofessional and there are plenty of other image sources online that provide a more polished finish. Try Flickr for copyright-free images although be sure to check the licensing permissions before you use them in corporate presentations.

- Get some books on preparing good slides, such as *slide:ology: The Art and Science of Creating Great Presentations* by Nancy Duarte. If nothing else, you'll pick up ideas on what makes an impact on audiences and what leaves them nonplussed.

TIPS FOR OTHER PROJECT COMMUNICATIONS

- Use video. This can be a good way to capture progress especially on building projects. Get a professional videographer involved at the beginning of the project and set out what you want and why you want it. Ask them for suggestions on how to record the evolution of a project visually. If your project doesn't have something that lends itself to being recorded under construction (such as software development), you can still use video. Put together a video montage with key stakeholders talking about the project, screenshots of the work to date and interviews with users. This can be a great way to spread the message and share progress on your project with the wider company at a staff conference or similar, and once the project is completed you could also share this publicly.
- Has a task been outstanding on your meeting minutes too long? Use your word processing package to make it flash when the document is viewed electronically. This type of humour can speed up results but can also backfire, so consider the person on the receiving end of the flashing task and be sensitive to how they will react.
- Sending a status report to someone who knows nothing about the project? Check it for jargon but also tone. Replace passive sentences with active ones: 'the system testing was carried out by the German IT team' becomes 'our German IT colleagues carried out the system testing.'
- Use photography. Take photos of your team at work, at key milestone dates or of users interacting with the end product. These can be far more powerful in sharing progress than words on a page.
- Make use of other forms of technology such as blogging to record progress on your project. A blog can make a good project diary, telling the story of your project as it moves forward.
- Try graphic recording, or at least using images in your team meeting to illustrate objectives. If you don't feel comfortable doing this yourself you can hire a graphic recorder to make a visual record of your workshop as this will often provide a richer source of information than minutes alone.

With any type of communication, set yourself challenges. Can you halve the number of slides? Can you shorten the report by two pages? If you print the update on both sides of the paper and make a booklet can you get it all on one sheet? Keep it short and simple; include the information the recipient needs to know, not a list of everything the team has done since the last update and above all remember who will be receiving the information and tailor it accordingly.

> Consider your audience when giving project updates and brush up your presentation skills: see the further reading section for some ideas of where to go next.

36 ORGANISE A PARTY

One of your roles as a project manager is to keep the team motivated throughout the project, so that they continue to meet deadlines and turn in quality work. The end of a project, after your final team meeting, can seem like an anti-climax to those people who have been working on it for months, sometimes years. They will quickly forget any little treats or motivational rewards that you organised during the project. Do something else to mark the end.

It is important to give the end of your project a sense of finality, especially if your project got canned. Team members are likely to feel that they have wasted their time and that their efforts went unnoticed. The official end of a project, whether it ran its natural course or was terminated early, is a good time to recognise everyone's commitment and achievement and to thank them properly for their work – what better way to do that than have a 'it's a wrap' party: movie makers have been doing it for years.

> Sam, a project manager in a large electrical engineering firm, had spent 18 months leading a 14 strong team of business and technical experts in delivering a new database system. The database allowed clients to look up the status of their orders online and within a few weeks of launch was already making a difference to the number of queries received by the customer service teams. The project got a small write-up in the internal magazine, but Sam knew that her close-knit, dedicated team – six of whom had given up many Saturdays and evenings testing the database – did not really feel that it was over. They needed to close the piece of work mentally as well as on paper, so Sam approached the company's HR department for advice. She found out that the firm had a central budget for rewarding high-performing teams and got an application form. Supported by her project sponsor, Sam completed and submitted the form. Two weeks later, she had authority to spend £30 per head on an event for her team, with the bill going to HR. She negotiated a deal with a local restaurant and the 15 of them went out for dinner one evening after work, as a fitting end to a successful implementation.

A formal recognition scheme, where such schemes exist, is a great way to offer your team a celebration for a job well done and also provide a firm feeling of closure. However, even if your company does have a recognition scheme you might not be successful in applying for funds. For example, your project may have been closed down prematurely or the funds may be very limited and you might be unlucky. However, you can still achieve the same result without spending a lot.

- Ask your project sponsor if they can provide a few bottles of wine or soft drinks for an after-work reception. The sponsor could also say a few words of thanks.
- Organise a pot luck picnic, where each team member contributes an item of food or drink and take a long lunch break in a local park.
- Does anyone in the team have useful contacts? Somebody with a large garden may be prepared to host a barbecue.
- Ask whether your local wine shop could arrange a tasting.
- What awards are available to enter? It could take several months for the judging to be complete and of course you could lose, so you could do something else as well in the meantime. However, awards are great for the CV and promote a real sense of achievement. The trade press will have details of annual awards relevant to your industry. Keep an eye on the local press too for news about local competitions.
- Ask the CEO or another senior executive to send letters to each of the team members thanking them for their contribution to the successful delivery of the project – even if you have to draft the letters yourself.
- If all else fails, get the team together for a group photo and circulate copies to everyone with a personal thank you note from you summarising the main successes. The notes will mean more if you tailor them specifically to the addressee instead of sending out a circular memo. If you are pressed for time or words, a printed memo to everyone with a handwritten 'thanks for all your hard work' on the bottom will just about do it.

However you decide to thank your team, make sure everyone can take part. A night down the pub may not be the right event for a multi-cultural team. One project manager organised a (rather ambitious) tank driving event before realising that his disabled colleague, who had been a key player in the project's delivery, would not be able to take part.

Closure of the project is important, but so is closure of the team as a unit. Mark the delivery or closure of the project with a notable moment to both provide a sense of 'ending' and to celebrate your work together, even if the project did not make it to the final implementation.

37 MAKE FRIENDS WITH THE PMO

Project Management Offices (PMOs) help project managers deliver projects on time and more effectively. However, PMOs continue to be challenged about the value they provide. A 2012 study by ESI International showed that while 72 per cent of companies had PMOs, over the previous 12 months 57 per cent of project managers not managed directly by the PMO had challenged the value that their PMO offered them. The study reported that only 18 per cent of respondents felt that their PMO had reached the final level of maturity, so there is still some way to go in improving the way that the PMO works with project teams and demonstrates its value to the company more generally.

Whatever the maturity level of your PMO, if you have one, learn to lean on them. They can be a great source of information and support.

DRIVING THE PMO

'The question "What does a PMO do?" is interesting!' says Gemma Viles, senior programme analyst in an IT PMO. 'Having worked at all levels of PMO from project up to strategic level in a number of organisations I can honestly say that the PMO job description has been different each time. I think the basic mantra for any PMO is first of all to provide help, then advice, experience, governance and anything else you can think of!'

Gemma often finds herself as a shoulder to cry on and a sounding board, as well as a soft skills coach, such as being asked for information about how to handle a particular stakeholder. She also provides practical guidance on templates and processes.

'I see the PMO as the glue that holds the doers to the thinkers,' Gemma explains. 'A PMO ideally speaks both languages, that of the project/programme and portfolio office and that of the business – financial, strategic, organisational goal-focused language.' Gemma's PMO has been fulfilling this role since 2010 and was originally staffed by a PMO manager and a programme analyst. Now the PMO employs a couple of contractors working the larger programmes to help with the day-to-day support. 'Everything else is up to the two of us,' says Gemma, 'and we really don't have enough time!'

Gemma thinks of the PMO as a profession in its own right. 'I've been a project manager and the "control freak" personality that's required for a good project manager is very different to that of a PMO,' she says. 'Although it's true that many project managers start off in support roles and that the PMO has always been seen as a stepping stone for greater things (i.e. project management) it's good to see PMOs coming of age and stepping out as a specialism of their own.'

Gemma's next challenge is to implement her PMOmid – PMO knowledge pyramid – model. 'PMO analysts make up the bottom tier whereby they help project managers a on a day-to-day basis, pushing out best practice, support and so on to the projects,' she explains. 'The next tier up is where the senior PMO analysts would sit, supporting programmes and higher risk projects. The top layer is for the strategic analysts, supporting large corporate, strategic and organisational change programmes. This layer would also look at corporate planning and would identify strategy. The strategy would then feed down the pyramid ensuring that all projects were aligned to the organisational goals.'

This model helps project managers know how their projects contribute to organisational strategy. Whether project managers are highly skilled in project delivery or highly knowledgeable about the business area the project change resides in, the PMO structure ensures they can deliver a successful outcome.

Gemma is familiar with the challenges PMOs face in proving their worth. 'If you take the PMO out you probably won't immediately notice the difference, until you realise your project managers are running wild and you've lost a grip on your understanding of both your portfolio and its progress,' she says. 'Projects are becoming more of our day-to-day bread-and-butter work as innovation and drive to beat the competitor mean we have to deliver bigger projects faster. Project managers may be the engine in the car to do this but the road, the highway code, the service stations and the emergency services are the PMO, supporting the project manager in getting to their destination!'

'The PMO strives to introduce and standardise economies of repetition in the execution of projects and is the source of documentation, guidance and metrics on the practice of project management and project execution,' writes Peter Taylor in his book, *Leading Successful PMOs*.[69] He goes on to finish the definition: 'It is also the body that links business strategy to the projects that such strategies require.'

The PMO can offer you a wide range of services including:

- templates;
- standard processes;
- support with training and certification, including delivering courses and managing a library of relevant books;
- resource management and handling timesheets on your behalf;
- coaching or mentoring for you or team members;

- quality audits and peer reviews;
- management of collaboration software and set up of tools such as a project wiki;
- a link into the network of project managers in your organisation.

PMOs also undertake a wide range of tasks to support the organisation as a whole, such as project selection, portfolio analysis and reporting. In fact, the remit of a PMO can be as wide or as narrow as the PMO leader sees fit and PMOs can get involved with practically anything if it facilitates the smooth running of change in the organisation.

The first step towards being able to draw on the support of the PMO is to find out what services your organisation's PMO actually provides. You could find that the PMO team can save you a lot of time especially when it comes to producing project documentation, as they are likely to have good examples of every type of document, log and electronic template you would want to use.

> When you know what they can offer you, tap into the services of the PMO and reap the benefits!

38 BE A LEADER

You don't have to be in charge to be a leader. You can demonstrate leadership skills at any level.

LEADING FROM BELOW

In Thomas Juli's book, *Leadership Principles for Project Success*, he writes about being a quality manager on a six-month project to develop a new software product which the company required for regulatory purposes. The project delivery date was delayed three times, and while the team did eventually deliver the software, it did not include all the original requirements.

This delay was down to the stakeholders having different views about the scope and the project manager failing to get the group to gain consensus. There was poor collaboration with a team, and a project manager, who were not committed full-time to the project. Attendance at project meetings was poor, and information sharing, especially about risks and issues, was practically non-existent.

Juli was not in a position to make sweeping changes to the project but he did demonstrate 'leading from below'. He worked with the project manager to tune the project objectives. He prepared project policies and procedures. He started his own crusade to improve communication through an online 'team room' and found that when he shared activities and information transparently, others joined in.

While all of this made a difference to team morale and no doubt contributed to the project achieving something, it was a 'truly frustrating experience'.[70] However, Juli recognises that it was an opportunity to learn – if only to know what not to do in the future.

If management is doing the right thing, leadership is doing things right. It is a skill that involves more than just technical savvy with project management techniques or software. It involves:

- setting a vision for the project with clear goals;
- communicating the vision and goals;
- enabling and motivating the team to achieve the vision and goals;

- evaluating progress and providing feedback as required;
- listening to and acting on feedback provided to you, and;
- removing roadblocks and making it easy for other people to do their jobs.

Being a leader involves maintaining an attitude that is open, honest, trustworthy and demonstrates integrity. 'Doing' leadership, however, is not something that can be codified and you will find yourself adapting your behaviours to suit different situations. What works day-to-day is not necessarily going to work during a crisis.

Day-to-day project leadership can be more hands-off. The team can have the latitude to do what needs to be done. You are steering the ship, but the crew know the course and are willing and able to help you get there. In a crisis, project leadership needs to be more directional. The team may not be willing to do what is required to put the project back on track, or they may not have the skills to do so. They will be looking to you as the project manager for direction and support. You'll need a more hands-on approach, reviewing every aspect of the project to find out where the project is not aligned – and then putting it right.

If you think that sounds more like the job of management, you would be right. Project recovery exercises involve leadership, but they are also very much concerned with management tasks. After all, leadership isn't something you can simply switch on when you feel like being a 'leader'. Leadership has to be part of how you perform your management tasks.

Carrying out the management of projects without that work being aligned to your leadership behaviours means that the project team will soon realise the lack of connection between the way you present the overall vision and the way in which you execute it. The management tasks – building plans, running meetings, handling change and so on – also have to be executed with the same approach.

> There is an overlap between management and leadership, and you should aim to develop a leadership attitude and set of behaviours that runs through both activities, predicated on openness, honesty, trustworthiness and integrity.

39 MANAGE A MATRIX ENVIRONMENT

The command and control approaches to management have largely died out in modern businesses and matrix management is the way things get done in workplaces today. Matrix management is where you do not have direct line management responsibility for the resources working on your project. Instead, you draw on a wide range of internal teams, outsourcing agreements and joint ventures to bring resources to your project. In this sort of environment, influencing and negotiating are key to getting things done, and trust becomes a very important currency.

MANAGING THE MATRIX

'We are very matrixed and rely on collaboration and project teams without formal reporting lines to get pan-company projects done,' says Monica Swanson, Director at a large global corporation. 'Due to our global environment, we are also very reliant on virtual collaboration as many times we do not get to ever meet team mates in person.' With around 60,000 employees in over 100 countries, project teams at Monica's company typically work with colleagues around the globe and with people representing various businesses. Ensuring that matrixed and informal reporting lines work well in this type of environment can be a challenge, especially when it comes to standardising information about projects.

Using professional project management software in this environment can be difficult too, due to limitations with licences and a user interface that can be confusing for stakeholders who are not used to working with Gantt charts or automated project reporting.

'To overcome this, we created a simple project tracking template in a spreadsheet,' explains Monica. 'A key to our success was a tab to track project details, and a secondary tab which auto-displayed current project status in the form of a summary pie chart and statistics. The summary tab was a great project view for stakeholders as they didn't need to see the detail. The detailed tab was great to track ongoing project details, and having the summary information auto-updated via formulae.'

This one-document approach provides a consistent way of managing individual projects, but can also feed into structured reporting across projects. 'We can take snapshots from each project template and collate them together,' Monica says. 'For example, we have a monthly status report which collates a project update (red/yellow/green status, and so on) for all of our current projects, and we distribute it to our key stakeholders like department heads.'

> All projects are tracked using the same template so all project teams (and their stakeholders) have consistency in how they receive information about the projects they are involved with. This ensures that when a team member or stakeholder is involved in multiple projects they don't have to get used to multiple ways of displaying similar information and provides standardisation across a matrixed environment. 'This uniform view into all projects makes for easier acclimation on a project,' Monica says. 'It allows people to join or form new project teams and have a standard means of understanding as we are all already familiar with the standard template. The template also has some standard fields which encourages best practices into all projects. For example, some people didn't think to allow sufficient time for test, and now this is more thought through as "testing" is incorporated into a drop-down list.'
>
> Standardised reporting across all projects is less time consuming for the team as they have a consistent snapshot to grab from each project. Additionally, executives get consistent information for each project, wherever they are in the matrixed environment.
>
> However, the individual project templates don't force linking between projects. This requires a degree of human intervention to ensure that across the matrix, teams and projects stay joined up. 'The fact that we have a standard monthly project template published keeps everyone informed of all projects,' Monica says. 'This allows all of us to stay better informed and make the connections as we read the project updates. We don't want all our projects to move forward in a silo, and thus our reporting helps connect the dots between projects and people.'

'Managers who struggle to delegate because they believe they are perfectionists, or lack confidence to really empower others, struggle to operate in virtual teams,' write Susan Bloch and Philip Whiteley, in their book, *How to Manage in a Flat World*. 'For many, who have worked in very structured hierarchical organisations, this transition to a different style of management often goes unacknowledged by team leaders.'[71]

While this is a generalisation, collaboration skills often come more easily to younger generations who have not been exposed to the traditional hierarchical environments of their older colleagues. Younger project team members will have learnt these skills through online communication and collaboration with their peers and may consider this way the only way to do good business. It is not, of course, the only way to get things done, and other cultures around the world continue to successfully operate with more formal organisational structures.

Projects, being temporary in nature, lend themselves particularly well to matrix management because you only draw in the resources you need for the time period in which they are needed. When the project is over, the team is disbanded.

However, matrixed teams can be a source of stress for project managers and they require great relationship management skills to work effectively. If you find yourself managing a matrixed team:

- Don't pull rank. The team members don't work directly for you, so they won't automatically feel obliged to do anything you tell them.
- Do keep their managers informed, and expect the same level of courtesy back from their managers when it comes to informing you of any holiday time that the team member has booked, for example.
- Do let team members have time to go back to their 'parent' team for team meetings and away days.
- Do rely on your negotiating and influencing skills.
- Do treat them as part of your team, even if they don't work for you.
- Do provide feedback to their manager (if requested) for end of year reviews or appraisals.
- Do work to build trust with your team members as soon as you can, through doing what you say you will and following through on your actions.
- Do standardise as much as you can, like reporting, templates, software and so on. This means that when someone is co-opted on to one project they have a similar experience if they are co-opted on to another at a later date.

Matrixed teams are now very common in projects, so polish your negotiating and influencing skills to give yourself confidence when leading people who do not work directly for you.

FURTHER READING FOR THIS SECTION

Bradbury, A. (2000) *Successful Presentation Skills*, 2nd edition. Kogan Page, London.

Bridges, W. (2003) *Managing Transitions: Making the Most of Change*, 2nd Edition. Da Capo Press, Cambridge.

Cook, S. (2001) 'Creating a high performance culture through effective feedback'. *Training Journal*, August 2001, 16–17.

Duarte, N. (2008) *slide:ology: The Art and Science of Creating Great Presentations*. O'Reilly, Sebastopol.

Kendrick, T. (2012) *Results without Authority: Controlling a Project When the Team Doesn't Report to You*, 2nd edition. Amacom, New York.

Madsen, S. (2012) *The Project Management Coaching Workbook: Six Steps to Unleashing your Potential*. Management Concepts, Tyson Corner.

Maggio, R. (2005) *The Art of Talking to Anyone: Essential People Skills for Success in Any Situation*. McGraw-Hill, New York.

Roam, D. (2012) *Blah Blah Blah : What to do When Words Don't Work*. Marshall Cavendish, London.

Turner, J. R. (ed.) (2003) *People in Project Management*. Gower, Aldershot.

West, D. (2010) *Project Sponsorship : An Essential Guide for those Sponsoring Projects within their Organizations*. Gower, Aldershot.

SECTION 4:
MANAGING PROJECT PLANS

INTRODUCTION

> *Moreover the hastening of any matter breeds disasters, whence great losses are wont to be produced; but in waiting there are many good things contained, as to which, if they do not appear to be good at first, yet one will find them to be so in course of time.*
>
> Herodotus (484 BC–c.425 BC), *The Histories (Volume II)*

Project planning is the process of identifying what tasks need to be done to complete the project and to meet the project's aims and objectives. Planning gives you, as the project manager, the opportunity to ratify with the stakeholders that you have really understood what they want, and that they understand what they are going to get.

Scheduling is the process of putting these tasks into the correct order. Schedules are flexible and it is up to you how you define 'correct'. Probably what is 'correct' when you start the project will not stay that way for long, and you will end up modifying your schedule as you go through the project.

Most projects are less than a year in duration and only 14 per cent of projects take longer than 18 months.[72] Even short projects need accurate plans and a managed schedule or you will find them taking a lot longer than was ever anticipated.

This section deals with planning, scheduling and time management in general, which are all key skills for project managers.

40 KEEP UP THE MOMENTUM

Starting up a project is often the easiest part. Keeping it going takes a lot of effort. Putting some thought into how you will keep the project's momentum while it is still in the honeymoon period will be time well spent.

THE PUST PROJECT

PUST stands for Polisens Utredningsstöd (police investigation support), and was a Swedish project to give police officers laptops on the beat so that they could cut down the amount of paperwork and processing of crimes that they had to do.

The project launched in 2009 with a team of 10 people and by 2011 had grown to 60 people, delivering system improvements every couple of months using an Agile approach. The rationale behind this was to avoid 'big bang' style releases, so the new software and laptops were rolled out to a region at a time, and also phased by type of crime, starting with possession of knives and drink driving and later adding more crime types.

While this phased approach allowed the team some early wins and proved that the solution was working, it wasn't all plain sailing. As the team grew, handoffs between the sub-teams became less effective. The team was eventually restructured with multi-functional sub-teams working on features, instead of split by role. This worked better, but one of the main areas where things still got stuck was system testing. The focus of the daily stand up meeting was changed: 'What can I do to contribute to system testing today?' This ensured that there was enough focus on keeping the momentum of the project going, as the bottleneck became a shared problem. One of the ways they worked to resolve it was to implement test automation. It was a culture change for the developers to stop working on new features for a while and instead start doing developments that would automate the repetitive work of testing. They also switched from a 'test at the end' model to 'test as you go' which sped up the learning of the developers so that they ended up writing better code. Showing everyone the bigger picture was a valuable tool in helping people to see that the project was more than just their part.

However, once they had removed the bottleneck of system testing they came up against another problem: changes were happening too quickly. The team were not used to dealing with this level of change, and neither were the users. The project manager introduced a low level of bureaucracy to control changes, requiring teams to do a short business case if a change affected more than just their area.

> Overall, the project was a success: crimes are now investigated 90 per cent faster than previously and a drink driving case now takes just three days to process instead of 31 days in 2010. The team know that there are more developments that they can do to provide even better investigative support resources to police officers in the field and will continue to release developments incrementally.[73]

Choosing to deliver incrementally is one of the main ways to keep the momentum going on a project. Can you implement some quick wins to show that the project is delivering benefits? This is also positive for team morale and senior stakeholders, as everyone can see that the project is making progress.

Relying on others is one of the main reasons projects falter. Other people have day jobs and priorities that will not necessarily align with those of the project. Consequently, a good project manager will be able to get them to do their part of the project without it being at the expense of their day job.

The trick is to make it easy for those outside your direct control to do their part, while also making it harder for them **not** to get involved. For example, if your team cannot follow the schedule you have produced in your planning software, transfer it to a bullet-pointed list. Ask them to tell you when they will complete their tasks: if **you** impose a date it is easy for them to say it was unachievable. It is much harder for someone to explain why it has not been possible to meet a self-imposed deadline. Communicate what others are doing and the benefits they are receiving. Offer as much help as you can while subtly increasing the peer pressure and finding answers to any excuses you hear as to why things are not progressing.

There is one big risk in situations where the momentum of the project is slacking because people are not doing what is required of them. Letting things slip over a prolonged period results in a stalled project which could be impossible to start up again (and the blame for its premature closure on your career record). 'Get it done fast,' write Robin Lissak and George Bailey in their book *A Thousand Tribes* about their experiences at PricewaterhouseCoopers:[74]

> 'In a large organisation, staying the course on any firmwide initiative requires speed – 30-, 60-, 90-day outputs – or it rarely reaches fruition. Unless the game plan is based on speed, a company tends to add time, effort, and bureaucracy to a project so that it never gets done.'

If you notice things starting to slow down, flag the deceleration to your sponsor as an issue.

 Be aware that any slowing down of activity could be the first sign of project demise so help your team to keep the pace by creating a structure where you can deliver incrementally.

41 PLAN FIRST – SET END DATE LATER

A project is always someone's idea, and that person always has an idea of when they want it completed by. One of the hardest things to do as a project manager is to manage the expectations of your sponsor, key stakeholders and team, who will all want to know when this project is going to finish before you are ready to tell them. A period of detailed planning at the beginning of the project is essential for two reasons:

- it provides clarity about what it is that you want to actually achieve, and;
- it gives you a firm foundation and confidence in your schedule dates.

After your planning activity is complete you can announce, with as few caveats as your risk management allows, the amount of time the project will take and therefore when it will be finished.

> **PLANNING FOR PUBLICATION**
>
> It's difficult to predict how long the publishing process will take when you don't know what or how many submissions you will receive. That's the dilemma faced by the editorial board of the international literary and arts journal *Upstairs at Duroc*. The planning process is dictated by the volume of work, which is only known after the submissions date passes. The magazine receives over 400 short stories and poems each year and once submissions are closed, the planning can begin in earnest. The editorial board can't confirm the publication date until they really have an idea of what selection process they have to go through.
>
> Two teams of dedicated volunteers read poetry and prose submissions and sift out those that do not make the grade. This process can take several months depending on the volumes. Then there is a second reading where the editorial board reviews all the remaining pieces over a number of weeks, makes detailed notes and then meets to agree the final selection. The prose team normally finishes first as the magazine receives fewer stories and the board often has very similar ideas about what makes a good piece of narrative. The poetry decisions are more challenging. Arranging the final editorial meeting is difficult as it needs to fit around the availability of board members, which can lead to more delays.
>
> Once this is done, the chosen pieces are handed to the layout team who prepare the magazine's 100 pages for the printer. A talented artist produces the front cover, and

> the magazine is set to go. The publication date can now be set, and the printer can confirm a date for delivery of the finished magazines.
>
> It isn't all guesswork: the team can use the schedule from previous years as an idea of how long things will take but can't assume it will be exactly the same. For example, they may need to commission new fiction or poetry if they don't receive enough quality submissions. They know that this approach to publishing is unconventional if you compare it to the way monthly magazines are produced, but *Upstairs at Duroc* prides itself on being an unconventional magazine.

To arrive at a position where you understand the overall duration of the project, start with the basics. Sit down with your team and work out the project plan, applying the same rigour and techniques to project planning as you do to the rest of your project activities. Take into account the advice in this book and best practice in your own organisation. What needs to happen? How long will each task take? The answers to these questions will allow you to pull together a timeline.

> In all but the most exceptional of circumstances, try to use estimates gleaned from the people who will actually be doing the work. If you cannot get hold of them, or someone with a similar profile or background to them, make an educated, conservative guess. Educate yourself if need be: find out if a similar project has been done before. Can you speak to the manager? How long did that piece of work take to complete?

It is important to be realistic about what tasks need to be done and the length of time each task will take. In a 2002 survey, having a realistic schedule was identified by 78 per cent of project managers as critical to a project's outcome.[75]

> It is a fine balance between spending too long at this stage and meeting the expectations of your stakeholders: you cannot keep them waiting forever. Aim for a best estimate end date, and tell them that's all it is until you have a chance to plan in more detail.

> Provide a date range that reflects the uncertainty in your estimate instead of committing yourself to a fixed date. For example, 'It will take seven months, plus or minus 20 per cent.'

It is always easier to plan the early stages than the later ones. You know what you need to do now in order to get things moving in the short term, but predicting what activities your team will need to do six months down the line or even later (with all the changes and modifications that come with managing a project) is much harder, to the point of sometimes being impossible. It is still possible to plan at a high level. Breaking your project down into phases or stages will help. Plan the first phase in detail and have a high-level plan with perhaps just a few important milestones for the subsequent phases. Phase One should include the planning tasks for Phase Two and so on. Your high-level plan will have enough details to satisfy a sponsor and give an overall impression of when the project will deliver, but be qualified by the need to do more detailed planning later. As the plan for each subsequent phase should be signed off by your sponsor, they will be obliged to agree and understand the flexibility within the end date.

There will be occasions when someone on the way to a senior management meeting stops you in the corridor and asks you to provide a date on the spot, before your planning is complete. If you truly cannot get out of answering, be as vague as possible. Say 'by the autumn' or 'during quarter one next year'. If pressed use your best guess as of that moment, plus some extra time as a safety margin. It is nearly always easier for stakeholders to manage the communications around a project that delivers early than for them to explain why it is late. The company's management team will not know (and will not want to know) the details of your project so they have to take your word on the duration.

Don't get sucked into the trap of promising an end date before you have really worked out the tasks involved. If you do have to provide an estimated date, try to offer a range or provide the level of confidence in the estimate along with the date.

42 MANAGE FIXED DATE PROJECTS CAREFULLY

Some projects are already time-bound when you receive them. Maybe there are regulatory requirements to meet by a certain date. Perhaps your CEO has promised a major customer something by the end of the year while they were out on a corporate golf day. Projects with fixed end dates present a different type of planning challenge for project managers. Instead of being able to analyse and plan, you are told what to do and when to do it by. Some degree of direction is good – after all, you cannot justify a three-month planning phase for a project that ends up only taking six weeks. The analysis part of planning has to take on a different spin when your implementation time is already ticking away.

When President John F. Kennedy announced in 1961 that the United States would put a man on the moon by the end of the decade many people thought he had over-promised. Instead, he sparked motivation and commitment which led to the successful Apollo 11 mission. Promising an end date upfront paid off for JFK but generally it is a risky strategy for managers: however fantastic your project, it is unlikely it will create as much loyalty and enthusiasm as a moon landing or have access to the same kind or volume of resources.

VALENTINE'S DEADLINE

Being part of a team staging a one-off bespoke party at a disused power station presented a challenge to the team at Gideon Reeling, an interactive experience company that specialises in performance pieces. The three-day Lost Lovers' Ball was scheduled for Valentine's Day weekend 2011, and had to happen on time. With a theme of 'the architecture of love', an exclusive supper club, live music and performance, and thousands of party goers expected in fancy dress, the brief was wildly creative. 'It took time to unpick and interpret the brief,' says Kate, one of Mr Reeling's assistants and performance artist. 'It was high concept and it was difficult to imagine how to prepare our sections creatively and immersively and relay a narrative in just three lots of ten minute sections that would also inform peoples' overall experience of the party.'

The Gideon Reeling team was responsible for production of a theatrical piece in various areas of the venue and some of those areas presented the need to adapt their production very carefully in order to adhere to the health and safety rules of the building. While it is an inspiring space, Battersea Power Station in London is derelict in parts and presented possible dangers to the attendees and specific constraints on use. Integrating a health and safety briefing to each group of audience without sacrificing the immersive nature of the show (by breaking out of character) and spoiling the theme of the ball presented a creative and functional challenge. Such considerations as enormous amounts of nesting pigeons, falling masonry, contaminated areas, dangerous drops as well as the presence of rats in the building all had to be taken into account.

'We wanted to use the exciting, hidden areas and those usually inaccessible to the public but discovered that all performers and audience needed PPE [personal protective equipment] to enter those areas,' Kate says. The team got round this by writing the requirement into the performance and adapting their narrative and characters to accommodate it which meant they were able to use the 1940s control room as a backdrop. 'The requirement to convey potentially clunky modern day health and safety information within our narrative, in fact, ultimately provided us with the key to resolving the narrative problems and interpreting the illusive brief. The result was a 1940s film noir style adventure into the dilapidated heart of Eros – led by a mysterious organisation – "The Trust",' Kate says. The final challenge was working with other contractors to try to tie them into the narrative to support the immersive nature of the whole party.

Despite a number of difficulties the performance piece and actors were ready on time. Kate believes that the key factors that made it a success were the ability to employ really good, adaptable people with a budget that enabled rehearsals on-site and a brave client who really trusted their abilities to deliver something exciting.

Watch out for these types of projects. Whatever the reason for the time constraint, the planning approach for time-bound projects should start off in exactly the same way as for those without time constraints. Work out the end date ignoring the fixed delivery date, even if you do have to crash your analysis time into a shorter period and use estimates with a greater degree of uncertainty than normal. If your schedule shows that you can deliver before the expected date, that's great. If not, move into proactive planning mode and get creative to find a way to deliver on time. Here are some suggestions.

- Run tasks in parallel instead of sequentially to save time. Look carefully at the critical path to find out how you could make yours work more efficiently.
- Who on the team is overstretched or overallocated and would more resources help? Ask for additional people to assist with the tasks where they would add the most benefit and help activities finish earlier.
- Work overtime, and get the team to do the same. Negotiate pay rates for overtime: don't be penalised for someone else's poor organisation.
- Postpone training courses or holidays, although be prepared to pay for this in the later stages of the project or during their next piece of work.

- Make decisions faster. Wheel out your sponsor earlier to speed up the decision-making process. Stop going to committees for decisions and become a 'just do it' team.
- Plan backwards from the end date. It gives you a different view of the project tasks and might make opportunities for time-saving more obvious.
- Cut out bureaucracy. If it normally takes a week to get a document signed off through the normal process, consider ways to speed this up, like walking round to each signatory's desk with a physical copy of the document and collect the signatures in a morning. Or call a meeting, go through the key points and have everyone approve it verbally.
- Look again at quality. Would the project be delivered faster if the quality criteria were lowered? For example, it might take a lot longer to design and build a website that has response times of less than a fraction of a second. If half a second response times were acceptable the website build would perhaps take less time. If necessary, another project or a newly formed Phase Two could improve on the response times later.
- Draw on your admin support team. If you're lucky enough to have a project support office, offload some of the administrative or routine project management tasks to them. You don't? Procure your department secretary or your boss's PA for jobs like being the minute-taker in your meetings and typing reports or even as an extra pair of hands for testing.
- Call in some favours to get things moving more quickly.
- And if all else fails, use your influencing skills to negotiate an extension for the delivery date.

Throwing additional resources at a project can speed it up, but beware the law of diminishing returns. 'The traditional method for varying the duration (and the cost) of an activity is through the allocation or removal of one or more resources for that activity,' write Richard Deckro and John Hebert in *Computers & Industrial Engineering*. '[This] is precisely the condition under which the principles governing diminishing returns may occur.'[76] There comes a point where allocating additional people to work on a task actually makes the output slower and you will have to identify this point. Extra resources can have a knock-on impact on the overall productivity as well. New people take time to get up to speed and someone has to help them during this period. It distracts the team from their own work, slowing down overall progress on the project.

It is rarely impossible to deliver on time, given the right amount of resources, an unlimited budget and a tightly controlled project scope, but projects seldom meet these criteria. With fixed date projects:

- plan creatively to slash time out of the schedule, and;
- get the support of your sponsor for when you have to steamroller through the office bureaucracy.

43 HAVE SHORT TASKS

Small projects have the added bonus of delivering benefits more quickly, thus ensuring political and financial commitment from your stakeholders for the next phase. If they can see immediate results, they are less likely to start thinking about the new solve-all project or divert your funds elsewhere. The same goes for short tasks within a project.

Short tasks provide a greater degree of visibility for progress, helping both the project's momentum and demonstrating to your stakeholders that the project is actually achieving something. The shorter the task, the easier it is to estimate accurately, meaning that you can have more confidence in your overall plans.

BREAKING IT DOWN

When Sharon Campbell organises a conference, she breaks down what needs to be done with the help of a project notebook. 'I have a page for speakers, a page for exhibitors, a page for publicity, and so on,' Sharon, a conference organiser for the South Eastern Colorado Chapter of the American Hearing Loss Association, explains. Conference planning involves multiple strands of activity. 'The venue is always the most critical,' she says. The association plans events for hard of hearing people and those who work with and care for them. As hearing loss is a factor in other medical conditions like brittle bone disease it is essential that the venue is equipped with wheelchair access, induction loops and other services. Sharon has links with several venues and tries to book early to make sure she has the choice of dates. 'I have to look at what else is happening,' she continues. 'We avoid Memorial Day weekend and Mothers' Day as turnout will be low, but we try to arrange things so that it falls within Better Speech and Hearing month.' One year the conference was organised the day after an assistive technology event finished so that educators and post-graduates could stay on for an extra night and gain continuing education credits by attending.

Planning the annual conference starts nearly a year in advance as Sharon, from Colorado, approaches exhibitors and sponsors to secure funding. 'The associations and companies we target plan their budgets at the end of the year for the forthcoming year,' she says. She starts contacting relevant groups in the autumn for the best response and documents everything in her notebook. 'You can carry paper around,' she adds, but also confesses to emailing herself copies of important documents to make sure everything is backed up.

> The publicity campaign breaks down into further pages of Sharon's notebook: print, TV, radio and leveraged publicity. 'Leveraged publicity is where we find groups with similar interests to ours and target them,' she says. 'For example, there are a large percentage of retired military personnel with hearing loss so I approach associations and magazines that specifically target that group.' By breaking down the organisation into smaller chunks, Sharon can keep on top of what was last said to which speaker and what magazine has agreed to carry which advertisement.

Aim to have activities with a duration of no longer than a week on your schedule. This makes it easier:

- for you to manage the plan;
- for you to spot slippage and monitor progress;
- for your team to manage their activity, and;
- for your team to estimate the length of their tasks.

While a week works well for many projects, you may find that it is too long for certain projects, especially those using Agile methods. Find a duration that works for you and the team and that makes it possible for you to track progress without having to get status reports every day if that isn't necessary.

Breaking down a large activity into smaller tasks encourages those responsible for the work to really reflect on what is involved. This should lead to more accurate estimates and will certainly make it easier for you to identify when an overall activity is going off track. You avoid the risk of a three-month task failing to deliver on time and needing another five weeks to catch up, delaying the start of other tasks and eating into any contingency you may have had in your plan. It is a lot easier to replan and redress any slippage when a particular five-day activity fails to complete on time.

> In large teams, those split over many locations or simply those where you do not have the opportunity to discuss the project on a regular basis, short tasks can help promote communication within the team. To monitor progress against the schedule you will, as a minimum, have to contact each person currently involved in a task to find out how it is going. With short tasks you get more opportunities to interact with the team and more opportunities to build a relationship with them. It is much harder to do this if you only have a conversation about project progress every six weeks.

Breaking down the activity into short tasks will give you potentially hundreds (if not thousands) of discrete pieces of activity. If you're building an apartment block you would end up with tasks like 'paint the walls of apartment one,' 'paint the walls of apartment two' and so on. Another option to consider is project segmentation, where repetitive similar activities are grouped together into segments that can be managed more or less independently. 'Network-based project management techniques [like critical path analysis, see Chapter 44] are difficult to use when most project activities are very long compared to the length of the project, and precedence relations among activities are not simple,' says Avraham Shrub, a professor at Tel Aviv University's Department of Industrial Engineering.[77] Building an apartment block is the type of project that might find network-based planning difficult because each apartment is worked on individually. The 'paint the walls' task will have a duration almost as long as the project itself.

> Project segmentation, according to Shrub is 'the division of the project work content into segments according to managerial considerations, as opposed to a division based on the type of work to be performed.'

Each department requires a similar group of activities performed in a similar order but each activity will be performed frequently and may run consecutively for each apartment or in parallel. Once the segments are defined the duration of each task and the relationship between the tasks can be defined too. You can manage the building of each apartment as a mini-project which gives you more flexibility in handling the plan and makes it easier to see progress at any given time.

Even projects that do not have repetitive elements that could be handled as segments can be broken down into sub-projects. Sub-projects have all the benefits of short tasks on a larger scale: even if the scope of sub-projects is mainly arbitrary and milestones purely administrative.

> Short tasks offer the opportunity to plan more accurately, show progress, gain commitment and deliver benefits.

44 UNDERSTAND THE CRITICAL PATH

The critical path is the longest route through a project schedule – the path that is made up of all the tasks which must be completed on time in order for the project to deliver to the planned end date. Critical path analysis works by identifying those tasks that must be done in sequence; those activities that cannot start until the preceding activity is completed. By default it also identifies those tasks **not** on the critical path where there is some leeway to delay work if you need to reallocate a resource to help catch up on lost time on a critical activity.

DESIGN TO DEADLINES

A project to launch a new product at Caroline Songhurst's food manufacturing company involves people from all over the company. 'My role is to co-ordinate the product design and to approve the final packaging,' she says. 'When we start a product launch the project manager explains the critical path, and we all know where we fit in.'

The critical path runs to a very tight deadline. 'Supermarkets are expecting deliveries on the day we tell them; it hurts our credibility if we don't deliver the goods, and leaves gaps on their shelves,' Caroline, a brand manager, explains. 'I work with design agencies and I can pass our internal deadlines on to them. If I don't hit my dates the packaging won't be ready on time and the project misses a critical milestone. I don't want to put my colleagues in a situation where they have to make up time because I didn't do something quickly enough, and I don't think the other team members want that for each other either.'

For each product launch the project manager explains to the departments involved how their role in the project affects the other teams. 'I can see how I fit into the whole project, and what my part means to the others,' Caroline says.

Knowing the critical path helps Caroline do her job because it gives her a view of the overall project and realistic target dates. 'The critical path for me is the most important part of what the project manager shares with us, apart from obviously the objectives of the product launch and what we actually need to do,' she says. 'The objectives for each launch are pretty similar, and I know my role in getting a new food to market. Those points are important, but for me the dates are the thing that decides how I organise my work.'

Critical path analysis results in a diagram representing the project. It starts in a similar way to producing a project schedule: with a list of tasks, a clear idea of how long they will take to complete and an appreciation of the order in which they need to be done. Table 44.1 shows a simplified list of the tasks required to start up a collection scheme for recycling glass.

Table 44.1 Task list for project to start up a collection scheme for recycling glass

Task number	Task description	Task dependant on:	Duration (days)
1	Sign contract with recycling plant	–	2
2	Buy lorry	1	5
3	Recruit driver	1	7
4	Distribute leaflets to residents	1	2
5	Distribute collection boxes to residents	1, 2, 4	3
6	Make first collection	3, 5	1

Figure 44.1 A critical path

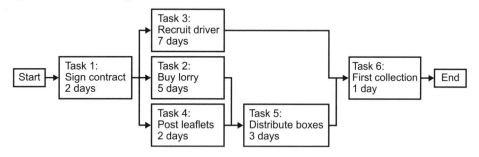

Figure 44.1 shows these tasks transformed into a critical path diagram, also called a network diagram. The boxes represent each activity and the arrows show how the tasks are linked together. Signing the contract is the first thing that needs to happen. After that, the project manager can start the process to buy a lorry and recruit a driver. Leaflets are already designed and will be sent out by post, so no need for the lorry to do that. Task 5 has to wait until the lorry is purchased, as it is needed to distribute the collection boxes, and also for residents to have received a leaflet explaining about the scheme. Once they have their boxes and the driver is hired, the glass collections can begin.

There is a number in each box referring to the number of days the task will take. Note that these are the duration, not the effort required. It will take someone two hours to stuff leaflets in envelopes and pop them in the post, but the elapsed time – real calendar time – between starting and having the leaflets arrive on doormats will be two days. It is these times that allow you to calculate the critical path, shown here with thicker lines. The critical path, being the longest route through the project with no breaks shows the earliest possible time the project can be finished: here it is Day 11. The diagram also shows non-critical tasks and where there is room for some movement. For example, if the recruiting process starts a day late, it won't matter as the collection cannot start until Day 11 anyway. But if the recruitment starts two days late this activity moves onto the critical path and the project now cannot finish until Day 12.

'It is often a surprise to find out what is critical and what is not,' writes Geoff Reiss, author of *Project Management Demystified: Today's Tools and Techniques*. 'Did you know, for example, that if you decided to build a tall office building, the critical path would normally run through the design, manufacture, and installation of the lifts?'[78]

One of the ways to save time on a project is to look at these tasks first. Adding more resources to critical tasks and getting them done faster will definitely have an impact on your end date. Tasks will step up and become critical though, so monitor the situation to ensure you are concentrating the team's efforts in the best place at all times. Planning software will calculate the critical path for you so with a click you can update task durations and work out what has become newly critical.

Critical path analysis is an aid to project planning that shows the tasks which must be completed and the order in which they must be done.

45 BASELINE YOUR SCHEDULE

You know that the schedule that you put together at the start of the project is going to bear little resemblance to what it looks like at the end. So what is the point of doing a schedule in the early stages of a project at all? Taking a snapshot at the beginning of your project is known as a baseline and can prove to be useful later on for many reasons.

BASELINING COHERENCE

David Schmaltz's position as founder of strategic project management consultancy True North has seen him work on many pressurised projects. One assignment, a hardware and software project for a high technology firm, found him working with, in his words, 'a dizzyingly broad community of contributors, suppliers, layers of arguing executive management, and customers'. With a seven-figure budget, and a global team numbering several dozen, significant effort was put into the planning in the early days.

The team covered a wall with sticky notes to create their plan to design and deliver the blueprint for a laptop computer. 'We were baselining coherence which allowed everyone in the community to think about the project in a similar way, whatever the plan and schedule might say.' David is clear that the project's original plan was the result of this effort to ensure everyone involved was talking about the same thing – or where they were not, that this potential contradiction was recognised.

A detailed project baseline was only drawn up for the foreseeable future, with the rest of the nine-month plan sketched in behind. The effect of the planning meetings early on ensured that everyone knew what was required to deliver the project successfully, and also who would be responsible for each critical delivery. 'Most of the planning energy was focused upon creating relationships between the people involved,' David says. 'We wanted to identify where critical relationships might exist and provide an opportunity for the people involved in these relationships to get to know each other, to get familiar with their working preferences, and to work together to achieve initial understanding and agreement.' He adds, 'By the end of the planning we were reasonably confident that everyone in the room was on a similar page. And where people were on different pages, we understood that, too. Everyone in the room, and later, everyone in the community, was clear that we had **not** created the roadmap we would follow to a successful conclusion.'

David firmly believes that the level of change involved in projects means they should not be a scripted performance but a fluid conversation between the stakeholders. 'The greatest danger any project faces at the beginning is mistaking unknowables for knowables, so we spent time identifying contradictions within the plan and almost no time attempting to resolve them into knowables,' he says. The team did not see any value in attempting to plan activities they did not as yet have all the details about. However, for the purpose of satisfying the funding authorities they prepared a summary including a list of proposed milestones with a caveat stating these would probably change, given the project's environment. In the event, the project stretched to 12 months.

The baseline plan was never formally captured and used to track changes, but in practice, much of it was not modified. The sticky notes were simply moved around as soon as a member of the team had better information about how a task would be achieved. 'This made the baseline a powerful medium for communication, rather than something that measured the goodness or badness of the effort,' David explains.

The focus of the project was not on picking up and focusing on why the hardware was not delivered to the original plan, but on the end goal itself. David was fortunate to be able to enthuse the senior management with this approach, but he acknowledges it was helped by the project's strong business case. 'I have not seen many projects that have been successfully managed by measuring deviation from initial baseline. Focusing on why it didn't turn out as planned can at best be a distraction from achieving the strategic purpose of the effort. This project, and many others like it, stayed focused on the prize.'

Setting a project baseline means taking a snapshot in time of your schedule. At the beginning of the project, all things being equal, this is how long you think the work will take. It's a target, a frozen view of a schedule which will no doubt change as the project unfolds.

Willie Herroelen and Roel Leus describe the baseline version of the schedule as a basis for planning external activities such as material procurement, preventative maintenance and delivery of orders to external or internal customers. Baseline schedules serve as a basis for communication and co-ordination with external entities in the company's inbound and outbound supply chain.'[79]

In another study by the pair they explain that 'booking' personnel to work on the project is another reason to have a realistic and stable baseline schedule.[80]

Your baseline schedule should be approved by your sponsor. However, this is just the first step. Just because you have a signed-off, baselined schedule does not mean that it is now set in stone. It would be unrealistic to expect it not to change at all. If you have discussed and agreed a time tolerance with your sponsor, it will not be necessary to go back for each tiny change provided you are within your agreed tolerance. For more on tolerances, see Chapter 4.

> Make sure your baseline plan is linked to a particular version of your project documents. If you're going to baseline your plan, baseline your other documents too. Your scope statement, design and requirements documents should all be under version control (see Chapter 19), so associating the relevant version to your baseline plan won't be difficult. This is useful because updates to the schedule normally result from the change management process. A new requirement, for example, will create a need to change the plans and have an impact on your schedule. When you look back during your post-project review you will be able to easily see how your baseline schedule was adapted following other changes.

Your baseline schedule provides a useful tracking mechanism so you can see how much has changed, which will help you monitor accurately. There are mathematical models for predicting the impact of change on your schedule[81] but it is more likely that you will use your time tolerance, contingency, the risk management process, as well as common sense, to accommodate changes to the schedule as your project progresses. The United States' Civil Engineering Research Foundation concluded in a study of Department of Energy projects that 'a rigorous risk assessment of alternative solutions under various scenarios provides a means of raising the confidence level that can be placed in early estimates'.[82] They also looked at rebaselining – that is, presenting a new, adapted (and normally longer) schedule to the sponsor and having that accepted and signed off as the new target timeline – and found that frequent rebaselining 'masked the true state of some projects'. Rebaselining is possible for your projects, but always keep a copy of the original schedule, however 'wrong' it has turned out to be. There are useful lessons to be learnt from why the original schedule was so inaccurate even if for project reporting purposes you now report against progress to the new schedule. This document will inform your post-project review and can be used to help improve estimating on projects in the future.

> Get your schedule approved and baselined at the beginning of the project as it will help you book resources and monitor progress.

46 MAKE MEETINGS PRODUCTIVE

Nobody wants to sit through a meeting and feel it was pointless, so why is up to half of the time attendees spend in meetings wasted?[83] It demoralises your team and is unproductive. Putting some thought into the objectives and format of your meeting in advance will allow you to get the most out of your attendees' time.

> **THE DAILY COCKTAIL PARTY**
>
> The Swedish police investigation support project, PUST (Polisens Utredningsstöd) ran along Agile lines and involved daily stand up meetings. However, with a team of around 60 people, the meetings became difficult to manage. The project team managed to get round this by splitting the meetings into productive, shorter sessions.
>
> First thing in the morning, the feature teams (made up of a requirements analyst, a tester and developers) met for 15 minutes. After this, there were synchronisation meetings of the specialist groups: testers from each team met, a developer from each team met and the requirements analysts met, again for 15 minutes. This enabled all the feature teams to synchronise their work effectively with the other teams. Finally, there was a third daily meeting, project synchronisation, in which a cross-team group discussed the whole project with the project manager and this was often attended by the project board.
>
> This may seem like a lot of meetings, but the whole thing was completed daily in 45 minutes enabling everyone to start the day knowing exactly what was being achieved. These meetings were called the 'daily cocktail party' with a focus on being upbeat.

The shorter the meeting, the more focused people have to be to get everything discussed in the session. This can really cut down on chit-chat and diversions. Another technique used often by Agile teams is to remove the chairs in the room – Agile meetings are called stand ups for a reason.

If your meetings are not creating the output you hoped for, you can cancel them; there is no requirement for a weekly project team meeting if you find that you are working closely with team members throughout the week and the formal meeting has become redundant. However, before you take that step, ask yourself why the time has become non-productive. If a meeting would truly be irrelevant to all the participants, then cancel it; do not waste people's time with a pointless get-together. But if you believe that arranging a time for everyone to be together would genuinely be beneficial if only you could increase the collective productivity, then think what you can do to make the meeting more efficient. If you would like to be able to squeeze something useful from the time, try some of these suggestions:

- Make sure people know why they are there, so they can plan their contribution in advance: send out an agenda and objectives for the meeting. In a research study carried out on engineering projects, Ana Garcia and her colleagues found that the agenda was the key to an effective meeting.[84] Allowing anyone to suggest agenda topics can mean that the final version of your agenda contains lots of items that are of no relevance to the majority of attendees. The researchers conclude:

> an ideal agenda contains only items that need the attention of mostly the entire group…purely informative items would be better dealt with through other means of communication. In addition, issues that concern only a few people in the group should also be discussed in another forum.

- Take it in turns to be the chair: this method also allows quieter members of the team an opportunity to speak and take the lead. However, keep the role of minute-taker yourself or delegate it to a trusted colleague, as you need to be confident in the completeness and accuracy of the minutes.
- Do not assume action points from the previous meeting have been completed – go through the minutes, ask for updates and if an action is not done, carry it forward.
- Set a firm start and end time, and consider imposing penalties for those who are late.

The guidelines above still apply for a meeting held 'virtually' i.e. by video, web or audio conferencing. The interpersonal relationships and reactions between your colleagues will be harder to understand and respond to if your participants are not co-located, so this method works better for short meetings with a clear structure and purpose. If you can't see who is speaking, ask everyone to introduce themselves when they start to speak ('It's Alan – I think we could deliver that feature by Friday.') You can also level the playing field by asking everyone to join by conference call even if some people are able to meet face to face and dial in as a group.

You can fit a lot into an hour-long project team meeting, if you plan in advance and brief your team as to what to expect. Plan for productivity and you will be surprised at the results.

47 DELEGATE SUB-PLANS TO WORKSTREAM LEADERS

On large projects your plan will include hundreds of tasks, if not thousands. Having all those in one schedule document is going to make tracking really tricky. You can get round this by delegating the management of sub-plans to your workstream leaders and only having high-level milestones in your overall project schedule.

When Claire Simpson was asked to run a project to refit a 60-foot corporate yacht, she knew she was looking at a schedule that would run to pages and pages of tasks. 'Our company specialises in refitting yachts, and from the work that needed to be done on this one I could tell it was going to be a tricky project anyway,' she says. 'I split the work up into chunks: there were tasks for the electricians, our IT technicians were involved for the onboard equipment, engineers needed to overhaul the mechanics and we were completely replacing the deck so that involved mainly carpenters, managed by our design architect. In all, I think there were about six different strands of work.'

Claire nominated a workstream leader to manage each strand, the senior team leader in each department, and asked them to prepare a plan. Once these were ratified and the dependencies between each plan agreed, she produced an overall project plan and associated schedule with everyone's key milestones on.

'Managing the plan like this was easier,' Claire says. 'We got together weekly to discuss progress, check where we were in terms of the client's schedule and just reassure ourselves that it was all going to plan. There are a lot of dependencies in yacht refits because the area we are working on is quite small and we can't have every team piling in at the same time.' Claire says she always works with workstream leaders, and it is a technique that helps manage contractors too. 'We don't have a permanent sail maker on staff, so if a yacht needs repairs or new sails, we contract the job out. One of the workstream leaders can manage that relationship,' she explains. What is important for her is to have absolute trust in the workstream leaders. 'We've been working together for years so I know that if they say a job is going to be done, it will be done,' she says. 'We can't do this job if we're not a team.'

DELEGATE SUB-PLANS TO WORKSTREAM LEADERS

A workstream leader is someone who manages a discrete group of tasks or people and reports to you. For example, you may have a project that involves input from the IT, marketing and customer service departments. Each department nominates a co-ordinator – a workstream leader – to organise the work their area needs to do. They can be a great help as they will know the people involved better than you and can select the right staff, as well as provide guidance on how long work going through their department will take, internal politics and generally be your expert in the field. An example of a project organisation structure is shown in Figure 47.1.

Figure 47.1 Example project organisation structure

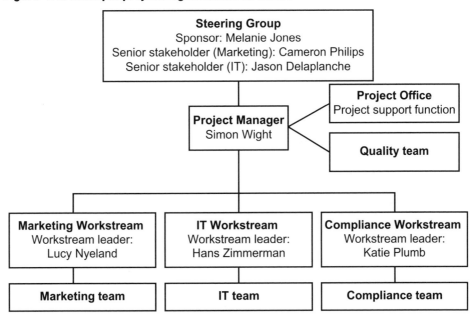

Once you are aware of the different tasks required to deliver your project, the workstream leaders can start to plan their areas in detail. Planning, in its widest sense, includes documenting all the deliverables and working out which order things go in. For their workstream plans, your workstream leaders probably won't feel the need to write down and describe anything in a very structured way, as they will have a clear idea by now of what you need them to do. You can encourage 'proper' documentation if your internal planning methodology requires it and you believe it would be useful for everyone, but the things that are really important for you are:

- the schedule – the list of dates saying what gets done when;
- resource usage to feed into your overall budget estimates;
- constraints that will affect the way they work;

- any assumptions they have made;
- the prerequisites they require before they can start, and;
- any dependencies their workstream has on other project activities.

You can then add any workstream level risks to your risk log. These details will also help you have a cohesive view of all the activity and how each workstream interacts with the others.

Whether your workstream leaders produce their schedule using formal project planning software or the back of an envelope doesn't really matter, as long as you both have confidence in the end result. Guide them if they have never produced a schedule before to make sure they include:

- start date;
- finish date (and therefore duration);
- name of person to do the work;
- dependencies between tasks within their workstream, and;
- dependencies on other workstreams.

Their workstream plans will include a lot of detail, much of it too low-level for you. That's not to say you shouldn't keep an eye on the detail, but delegating management of tasks that take just a few hours is a better use of your time, and will help you keep focused on the overall objectives.

Make your reporting easier by copying their major milestones and dependencies into your project schedule. When you need status reports, the workstream leader only has to tell you how far their team has got in relation to the milestones, and you can record the summary progress on your own schedule without ticking off thousands of tiny tasks. Don't go too long without an update. Depending on the type of project, two weeks is about the maximum you can leave it before getting a status report.

Delegate the management of sub-plans to workstream leaders to keep your overall project schedule a summary of all the activity across multiple teams.

48 MANAGE PROJECT DEPENDENCIES

A dependency is a relationship that links the order in which activities are carried out; Task B is said to be dependent on Task A if the start or finish date of Task A must be reached before Task B can start.

No project exists in a vacuum. Dependencies are links between tasks or projects, and you can be pretty sure that your project will have some. Dependencies form part of the environment in which your project will be operating. Understanding this environment will give your planning activity more clarity.

SLOWING DOWN THE RATE OF CHANGE

Rosa and Carole were managing separate IT projects at a healthcare company, both with the objective of installing new software. 'At one of our routine team meetings with all the project managers, we went through the high level milestones for everyone's projects,' explains Rosa. 'We realised that Carole's project and mine both had the requirement for staff in one business unit to learn how the new software worked. Unfortunately, the timings looked off and it would have meant the business team having to pick up both new applications at practically the same time.'

Rosa and Carole realised that this was an unrealistic expectation to have and that it would be better to amend the project delivery dates. 'There was a further complication in that my project was updating the operating system software to Windows 7 and installing Microsoft Office 2010, and Carole's was a finance business application,' Rosa continues. 'I needed to install my software first, so that Carole's would work effectively, and that would save her project team testing that the software they were installing worked on both the old and the new operating systems.'

By mapping the dependencies between the projects, Rosa and Carole worked out that it would save Carole's team time if they only had to test compatibility with the new operating system, which meant it made sense to speed up Rosa's project. They also agreed that they would wait three months between installing the new Office suite and the finance application, to give the business team time to get used to using one product before having to take on learning another. 'We didn't need to make the

163

> projects dependent on one another, but it saved us effort and it gave the business team a better service because they weren't bombarded with new software all at the same time,' Rosa says. 'Both projects had a better chance of succeeding because they had adequate business support.'

A project will be dependent on other activities happening within the company (outside-project, in-company) as well as on things happening outside the company and outside of the realms of the project (outside-project, outside-company). Many projects will also have in-project, outside-company dependencies in the form of third-party arrangements or reliance on contractors. Finally, there are also the dependencies within the project itself, namely that one activity cannot start until another is finished (in-project, in-company). These types of dependencies and some examples are summarised in Table 48.1.

Table 48.1 Types of dependency

		Company	
		In	**Out**
Project	In	Sequential or dependent project tasks, e.g. testing the IT system is dependent on first writing and building it	Contractor deliverables, ad agency copy, printing, outsourced IT deliverables
Project	Out	Linkages to work in other departments or on other projects, changes in company strategy or policy	Regulation, compliance requirements, health and safety standards, government or industry policy changes

Having identified your dependencies, work out their impact on your project. Some, especially the outside-project dependencies, will need to be entered into the risk log. You may intend to manage others through your plan, by keeping an eye on how the situation is developing. Include any relevant people in your communications and be certain to come back to your list of dependencies regularly and add in any others as they emerge.

> Your project will be dependent on not only your interpretation of the business situation at the starting point, but also on other vital factors. Identifying and understanding these will help you to plan and manage the project accurately.

49 MANAGE MULTIPLE PROJECTS AT THE SAME TIME

Unless you are working on one very big project, it's likely that you'll have multiple smaller projects on the go at the same time. Keeping track of what is happening on what project can be a challenge, as can balancing the priorities between them. Sometimes it is a case of working on the project for the sponsor who is shouting the loudest, but that's not a great way to prioritise your work.

JUGGLING TECHNIQUES

Gina George is the only project manager for her company's Call Centre/Public Safety division, which implements server-based recording, agent evaluation and workforce optimisation systems throughout North America. 'Although there are some seasonal fluctuations, I average about 15 active new or upgrade implementations at any given time,' she says. 'Another four to five projects are typically in some state of monitoring and can't be closed due to an issue like an unforeseen software problem.' In the first half of 2012, Gina managed 40 projects to closure, so she knows how to keep on top of multiple projects.

'When I took over this role I was able to put some systems in place that help me keep track of new projects coming in,' she explains. 'Nevertheless, managing multiple projects is a herculean task at times. I do pretty well when my project count is between 10 and 12, but on those occasions when it has crept closer to 18 to 20, I simply am not able to attend to every detail in the way I would like.'

Gina relies on automated calendar events and tasks to keep on top of things, and she also has great technical leads who can help pick up the slack. 'Administrative details are the most likely to be missed as I don't have good back-up there,' she says.

Excellent record keeping helps her keep track of progress on multiple projects. Gina begins each project by creating a project workbook in a spreadsheet package. This is uploaded to the Microsoft SharePoint site where the working copy is maintained for the life of the project.

'Because we're not yet a paperless company, I also keep a project folder,' Gina explains. 'While the project is in progress, these folders reside in my office and each has a post-it attached to the front with key project information, milestones and completion dates. At the project's end, this converts into a permanent folder of customer information.'

> While Gina's systems help her track her projects effectively, there are occasionally times where something falls through the cracks. 'In the not so recent past, I had a technician fly into a site only to find that the server had never been shipped from our office,' Gina confesses. 'The error occurred because I skipped a checklist before going out of town for training. The short answer to what I'd have done differently is not skip the checklist! But what I'd prefer to do differently is have an administrative support person who can handle this type of detail on a continuing basis. Unfortunately, we are the typical "do more with less" company so that's likely not in the cards.'
>
> Gina finds it interesting that more and more advertisements for project managers stress the need to manage multiple projects, while a growing body of evidence suggests that multi-tasking is not the most efficient way to work. 'I wonder if these companies realise that managing a single project requires significant multi-tasking,' she says. 'There comes a point when the multi-multi-tasking required to manage the number of projects I do will inevitably result in project failure, burnout, or both. Having said that, I love my job. I only want to love it more efficiently.'

Small projects have some characteristics in common. According to Sandra Rowe in her book *Project Management for Small Projects*,[85] small projects typically have:

- a duration of less than six months;
- fewer than 10 team members working on them on a part-time basis;
- a single objective with a clear scope and straightforward deliverables;
- a single decision maker, and;
- few interdependencies in skill areas.

That may sound relatively straightforward to manage, and by itself one small project is. However, when you have to manage multiple projects you start to hit additional challenges such as:

- Not having enough time to manage the workload effectively.
- Having to multi-task every day.
- Not being able to access the right resources in a timely fashion.
- Having to manage shared, common resources who are not dedicated to your project and who get pulled onto higher priority work at a moment's notice.
- Being on your own with little support and still expected to deliver to the agreed deadline.

If you have been in a situation where you have had to manage multiple, unrelated projects then you can probably think of other challenges too.

It's worth a quick look at the dangers of multi-tasking. Research from the University of California[86] shows that most people are interrupted regardless of what they are working on. When a task was interrupted, only 77 per cent of people went back to it that day, with the rest of the survey respondents leaving it for another day. This could have a massive impact on a project team member's ability to hit deadlines. The researchers also found that open plan office workers experienced more interruptions but they were able to work on one thing for longer before they switched to something else. This could be because in an open plan environment colleagues can hear and see when a natural break occurs so they take that moment to interrupt.

The researchers put forward three recommendations, which you could try in your workplace if you find yourself getting constantly interrupted and side-tracked on to other tasks. Ideally, interruptions should be about the work that you are currently doing (although they don't provide any suggestions about how to make sure this happens). It should be easy to switch between tasks, so make life easier for yourself with standardised project documentation, plans and reports and keep your records tidy. Finally, they recommend making it easy to go back to the task that was interrupted by preserving the state it was in so that you can quickly pick up where you left off. Software that auto-saves what you are working on is one way to do this, but you could also use the trick of stopping your email or document mid-sentence as this will prompt your brain to pick up the train of thought more quickly than if you complete an entire paragraph.

When faced with managing multiple projects you will have to try to avoid the day-to-day challenge of multi-tasking. However, there are some things that you can do to make multiple projects easier to manage.

1. **Categorise your projects**
   Try to find some similarities between projects. Group them by sponsoring division, by the type of resource working on them, by client or by any other criteria that makes sense to you. This will give you at least some logical way of grouping your work so that you can focus on groups at a time and avoid shifting between radically different tasks.

2. **Prioritise your projects**
   If you are not given direction about what project should take priority, you can work this out yourself based on deadline, benefits, the availability of key resources or simply which project sponsor shouts the loudest. This will give you some structure to your work so you know what you should be focusing on first.

3. **Apply consistent management techniques**
   Use the same methods and tools for each project. Standardised templates for the project charter, schedule, risk log and so on will save you time and also avoid some of the mental jumps between tasks when you inevitably have to switch. At least when you move to working on a different project the structure and documentation will look the same.

4. **Manage across projects as well as individually**
   Once all your individual projects have plans, consolidate these into a large multi-project plan. You can do this on a spreadsheet or by rolling up schedules in a project management tool. Use this multi-project view to manage risk across your workload and to help you plan the best use of resources. Essentially, this is portfolio management so it's a good discipline to practise!

Group your work when you are managing multiple projects and put in place strategies to help you stay on top of all the tasks. Avoid multi-tasking if you can.

FURTHER READING FOR THIS SECTION

Berkun, S. (2008) *Making Things Happen: Mastering Project Management*. O'Reilly, Sebastopol.

Calhoun, K. M., Deckro, R. F., Moore, J. T., Chrissis, J. W. and Van Hove, J. C. (2002) 'Planning and re-planning in project and production scheduling'. *Omega*, 30, 155–170.

Campbell, C. A. (2007) *The One-Page Project Manager: Communicate and Manage any Project with a Single Sheet of Paper*. Wiley, Hoboken.

Devaux, S. (1999) *Total Project Control: A Manager's Guide to Integrated Project Planning, Measuring and Tracking*. Wiley, New York.

Gardiner, P. D. and Ritchie, J. M. (1999) 'Project planning in a virtual world: information management metamorphosis or technology going too far?' *International Journal of Information Management* 19, 485–494.

Harrin, E. (2010) *Social Media for Project Managers*. PMI, Newtown Square.

SECTION 5: MANAGING YOURSELF

INTRODUCTION

> *Personal influence is not to be trusted beyond a certain limit.*
>
> Caius Cornelius Tacitus (c.56–c117), *The Histories, Volume II*

Researchers can't agree about what makes a good project manager. Each new survey produces a different list of the top 10 characteristics for successful project managers, which just goes to show that a 'model' project manager does not exist. Every individual approaches project management in a different way, bringing with them a unique set of skills and experiences that they devote to getting the job done in the best way they know how.

This section cannot teach you how to be 'better' but it can give you some ideas for developing yourself, especially in time management and career progression. It also presents some examples of when it all goes wrong: project managers in difficult situations and how they rose to the challenge.

50 GET ORGANISED

It is a project manager's job to organise everyone else, and you will be much more efficient at doing that if you can keep on top of your own activities. If you are clear about what you have to do next it will make it easier for you to organise other people and the work of your team.

ON TOP OF THE LIST

Former corporate finance executive, Tina, from Montana, found the best way to manage her daily activity was to keep a list of all the tasks she needed to complete. 'I kept my list on a plain legal pad. A lot of my co-workers were really getting into computer organisation software, but they were always tinkering and it took them 10 minutes to get the thing to record the task. My paper method would require me to "rewrite" the list on a clean sheet every once in a while as I got through stuff and added things.' Tina found a use for all the old lists as well. 'I kept all my old to do lists in a file in my desk for use when I prepared for my annual review,' she says. 'After a year, who can remember all the little things they've done that ended up making a big difference? Keeping those old to do lists was a **huge** help, especially when asking for a raise.'

Tina has some advice for those who want to use this method: 'The importance of the list is keeping **one** list – not a bunch of sticky notes on your monitor, not a thousand little scraps of paper. If I had to remember to make a phone call, it was on the list. Meet a deadline, mail something, call the doctor before 5 p.m., get back to someone, whatever,' she says. 'Giant projects sat on the list, like "Corporate Budget", until the project was undertaken, and then I would break it out into the thousand pieces that had to be accomplished.'

She found a great deal of satisfaction in drawing a line through a completed task, and realised that her to do list also provided an easy way to demonstrate to her managers what it was she spent her days doing. 'The List is a great way to show or, for some people, prove, how busy you really are,' she says. 'There is a lot of political capital gained with your boss when he or she sees you pull out The List to discuss what you're working on, especially if The List is long.' Tina is sure that having a list of activities helped focus her attention and make her more productive. 'The List Game became an endless cycle of me trying to find a way to get everything off the list,' she confesses.

Keeping yourself organised is a step towards being more effective and being able to organise the others in your team. If a paper list to organise your activities is not going to work for you, experiment with some other methods until you find one that does:

- Have a separate section on your project plan for 'project management tasks' and use that to keep on top of your activity.
- Use a To Do list application on your phone or tablet.
- Use the back page of your project notebook to record your list. Mark actions with an Ⓐ next to them when taking notes in a meeting, and then transfer them to your list at the end of the day.
- Use the task list feature in your email programme.
- If you can hold the list in your head, practise mentally juggling the order until the most important items are at the top without forgetting the smaller jobs.

Knowing **what** to do is only half the challenge. Knowing **when** to do it is almost a separate skill. Being able to prioritise your activity is another trick to get the most out of your time. Take your list of tasks and work out for each one its relative importance and urgency in relation to the others. Some tasks can be incredibly important, but not very urgent – at least for today. Figure 50.1 shows the four categories of prioritisation. Both identifying tasks and prioritising them are things to be done on a very regular basis, as projects shift on a regular basis and what is urgently needed for tomorrow may suddenly become less urgent giving you more time to focus on something else.

If you are struggling to keep on top of your list of things to do, and find yourself with not enough time in the day, make your phone calls first. A call before coffee break in the morning will not take as long as the same conversation at the end of the day. In the morning, everyone is busy sorting out their day and will want to get the conversation over. At 4.30 p.m. people are starting to look forward to going home and will drag the conversation out.

Figure 50.1 Urgent and important tasks

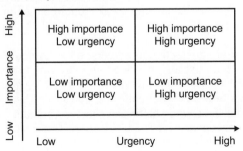

One of the crucial skills of a project manager is organisational ability. Find a way to keep on top of the tasks that need to be done that works for you.

51 DON'T DO ZOMBIE PROJECT MANAGEMENT

One of the first things that project managers learn on training courses is that a project has a start, a middle and an end. Unfortunately, some projects outside the classroom don't exhibit those characteristics.

KILLING THE ZOMBIES

'A little over a year and a half ago, I left my position as an advisor to a US Government PMO,' says Derek Huether, an enterprise Agile coach and trainer. 'It was a traditional PMO, filled with certified professionals doing very good work in very specific areas of expertise. We had budget specialists, schedule specialists, earned value specialists, and even contract specialists. The projects being managed were very traditional in nature, with the exception of one thing. They never ended!'

This seemed odd to Derek, who had committed the definition of 'project' to memory when he was pursuing his project management certification. 'A project is a temporary endeavour with a defined beginning and end,' he says.

He might have been able to forgive the long timescales if this PMO was managing projects committed to the construction of battleships or the next generation of an interstate highway system. 'I realise these would take years to plan and implement,' he says. 'But this PMO was managing software applications, both in development and in operation. So, why is it these projects never ended and never seemed to deliver anything of value? Why is it they just wouldn't die? It is because, of course, these were zombie projects.'

Derek who is active in the PMI Agile Community of Practice and the author of *Zombie Project Management*, defines a zombie project as 'one that merely exists'. 'It consumes resources in the form of time, money, and lost opportunity,' he explains. 'Zombie projects exist for so long, people forget why they were created in the first place. You hear people say "that's what we've always done" and nobody asks why.'

Faced with a set of projects that were simply taking up space, Derek had to make a choice. 'I would like to say I killed off all of these zombie projects or at least put them all back on course to delivering value to the taxpayers,' he says. 'Alas, if you are a connoisseur of zombie movies like I am, you know there is only one thing you can do. Divide and conquer.'

> Derek identified one small project that he knew a small team could bring back from the brink. 'We identified what deliverables were most important to the project sponsor,' he says. 'Then we established a small cross-functional team, asking some members to work outside of their functional expertise, to allow the flow of work to be regulated.' Derek took an Agile approach, scheduling a daily, 15-minute standing meeting, to communicate status and deal with issues and risks. 'We asked the team to demonstrate to the project sponsor what they had got done in the previous week, to allow the sponsor to accept or reject what was done and provide feedback and insights into shifting priorities,' he says.
>
> Derek and his team established regular weekly deliveries of product, to build onto previously delivered items, and they clearly identified a project end date. 'Whatever we had done by that date would be it, without new project sponsorship and the creation of a new project following the same parameters of our new non-zombie project,' he says. As a result, he brought this particular zombie project back under control and ensured it delivered something of value to the project sponsor.

Don't be the project manager that lets a project go on and on through accepting every scope change and picking up all the operational work as well as the requests for new functionality, additions or improvements. That doesn't make you a project manager – although you could find that there is a useful long-term product management role for you if you prefer this way of working.

On large programmes, projects and operational work coexist, so it is quite likely that at some point in your project management career you will be asked to get involved with operational work, and it may feel like your project is never-ending. However, zombie projects are not operational, day-to-day work. They are projects without structure, with poor management, poor oversight and lack of governance. Like the zombies in films, they are mindless. The project managers and the teams leading them don't ask the basic question – why? Why should we do it like this? Why are we doing this project at all? If these projects can't be rescued and turned into something useful, they should be closed down. After all, they are sucking up valuable resources that could be put to better use working on something else.

> Don't let your project turn into a zombie project – taking up resources and energy that could be better used elsewhere. You might not be able to stop it, but at the very least recognise when it has happened and recommend that the project is closed down. Otherwise you'll be mindlessly supporting a project that is going nowhere and has nothing to offer. Cut your losses and run!

52 KEEP YOUR RECORDS TIDY

Project management produces a lot of paperwork. As well as the official documentation, there are also meeting notes, reminders to yourself, scraps of paper where you have scribbled budget changes or useful telephone numbers...the list goes on. Not only will you have piles of paper, but you'll also have electronic notes stored on mobile devices, in emails, task management software or attached as virtual sticky notes to project plan or calendar entries. Keeping all this in a semblance of order is important, as it means you can instantly find everything you need.

UP IN SMOKE

Project manager Celine Montague was forced to review her approach to storing information when her PC suffered a terminal crash and she lost all her project documents, which literally went up in smoke. 'All my notes for our European fraud project were on my hard drive, and not backed up,' she says. A freak power surge at Celine's Brighton-based company put a handful of computers out of order. She realised what a mistake it had been to not have a back-up when a curl of grey smoke drifted out of her PC and had her racking her memory to recall critical project dates and decisions. 'Since then I've always kept backups and paper copies of really important documents,' Celine says. 'I learned the hard way!'

Project managers are largely left to their own devices when it comes to deciding on a method of organising their records. There are a myriad of ways of storing project information now and your Project Management Office may have an online document repository or software that it encourages you to use. The easiest option for you is to adopt what is currently in use by other project teams in the company, but if you have to create your own filing system from scratch, here are some tips.

KEEP THESE IN PAPER FORMAT:

- Signed copies of contracts including appendices, amendments and service level agreements.
- Any insurance policies you have set up as part of the project.
- Escrow agreements.
- Signed copies of change control notifications to contract terms.

KEEP THESE IN ELECTRONIC FORMAT:

- All the above – scanned in and stored electronically with the rest of your project files (as well as the paper copies, as these will be good for easy reference).
- Purchase orders, invoices and quotes.
- Contact details for the project team.
- Other project documents like the project charter and schedule.
- Risk, issue, change, dependency and action logs.

SHRED THESE:

- Paper copies of purchase orders and invoices – your corporate finance team responsible for Accounts Payable will also store copies, so you don't need these.
- Paper copies of quotes, once you have stored a copy electronically.
- Any other project documentation you print out – practically everything you work on will have some degree of company confidentiality, so don't put anything in the normal office waste paper bin.

BE ESPECIALLY CAREFUL WITH THESE:

- Financial records relating to project team members, such as salary, expenses and bonus arrangements.
- Performance appraisals or feedback forms for project team members.
- Timesheets.

Keep sensitive information somewhere separate from your main project filing.

Don't store information solely on your laptop or mobile device – back it up to a shared network drive or enterprise project management tool so that if your device gets lost or stolen you don't lose the project records as well.

Once settled on a routine – good or bad – it is easy to become stuck, finding workarounds to a poor or complex filing system because there is no time or inclination to properly review and revamp.

Your filing system needs to be something that works for you but that is also time efficient. Try a number of different techniques until you find something that works for you and that is compatible with how the rest of the organisation works.

> Keeping your project records tidy means it is easier for you to find information and also to be clear about what you are missing.

53 MANAGE YOUR PERSONAL BRAND

You are a highly competent, technically excellent project manager who delivers results time after time, so why are your efforts being ignored? Unfortunately, it is often not enough to do a good job, but you also have to make sure you are *seen* by project sponsors to be doing a good job. And not only are project sponsors frequently uninterested in your activities, they also have a very different idea to you as to what constitutes a job well done.

THE IMPORTANCE OF A PROFESSIONAL PROFILE

'Would you hire an individual dressed up in a bed sheet as a make-shift toga?' says Dr. Andrew Makar, an IT programme manager and the author of *Project Management Interview Questions Made Easy*. 'As much fun as attending a toga party could be, you don't want any questionable profile pictures appearing on Facebook, Twitter or any other social media platform. With social media platforms, you don't control the content or who has access to the latest picture you may have been tagged in. I've gotten into "trouble" with a few personal friendships when I didn't even post anything to Facebook, yet someone else did and that was all that was needed to spark controversy.'

According to Andrew, it is important to maintain a clean, professional project management profile because today your digital footprint becomes a part of your CV that can be difficult to erase. He has chosen to use Facebook in a personal capacity and LinkedIn professionally.

'My LinkedIn profile is professionally maintained and reflects the image of a project management professional,' he says. 'I'm a fan of World of Warcraft style role-playing games, but you won't see it on my LinkedIn profile. You can still maintain an engaging and professional online profile by sharing relevant news and constructively commenting on relevant status updates.'

Your online profile is one of the publically available pieces of information a hiring manager or an existing employer can find without your permission. Which image do you want them to see? A project management professional speaking at a local project management conference or a picture of you in a toga?

A quick search online will highlight a number of cases where individuals have missed out on job opportunities or lost their jobs because of what employers found out about them online. 'I haven't missed out on any opportunities,' says Andrew. 'However, promoting my blog on my LinkedIn profile did cause a few issues when one of my peers tried to damage my reputation at work.'

'Conduct a professional edit of your Facebook page, Facebook photos, LinkedIn profile, Pinterest, Twitter feeds and any other social media sites where you have a profile online,' recommends Andrew. 'Cleaning up old posts or removing questionable links is easy enough to do. Even if there are some questionable tweets or discussion forum posts from a decade ago, by creating new content that is relevant to your profession, it will suppress the older content. However, once content is indexed it is hard to delete it permanently. Be careful and professional!'

He recommends posting relevant but sanitised presentations on project management topics, and links to relevant project management articles. 'Be a thought leader without compromising your personal and professional profiles,' he says.

Promoting yourself in a professional way is a delicate balance between shouting about your own achievements and seeming arrogant, especially when you are working to improve your personal brand and reputation within your own company. 'Help your colleagues,' recommends Andrew. 'No one wants to hear you brag about your accomplishments. Rather than promote yourself through your accomplishment, promote yourself by helping others with their project management challenges.' If your company has an established community of practice, then this provides another forum to showcase your knowledge without having to appear a braggart.

'In a fluctuating job market, your personal brand is your safety net,' says Andrew. 'If you have a strong personal brand, it may help you with job stability within your company. Even if you lose your job, a strong personal brand will create new connections, contacts and opportunities that lead to new career opportunities.'

Lynn Crawford, a Sydney-based researcher, studied the gap between how good project managers were at doing their job and how good their supervisors thought they were.[87] She tested over 200 project managers against both their knowledge of techniques and their practical ability to do the job, using the Australian National Competency Standards for project management. Then she asked their supervisors to rate them on four criteria: their value to their clients and to their organisation, the ability to motivate others, and their ability to work with colleagues to deliver a project. The results were surprising. There was no statistical correlation between doing well on the tests and receiving a good supervisor score. Crawford concluded that 'the knowledge and practices valued by project management practitioners, and embodied in their professional standards, are not the same as the knowledge and practices valued by senior managers.' When she looked deeper into the results, she found that if project managers wanted to be perceived as excellent at work by their supervisors they should concentrate on:

- working on projects with high level of ambiguity in either scope or methodology;
- working on projects that differ greatly from each other;

- developing an excellent knowledge of cost, time, procurement and HR management and using these skills to control their projects, and;
- not becoming too involved with general management activities, to avoid being seen as interfering.

In an organisation where projects are allocated to you without the opportunity to choose or express a preference for your work, the first two recommendations may be difficult to achieve. Concentrate on trying to influence those people that make the decisions about the projects on which you work. If they see you starting to operate at a more senior level, there is more chance of them giving you the challenging projects you need, which will in turn raise your profile elsewhere in the organisation.

Following Crawford's last two recommendations will improve your skills and how others perceive you, but you will still need to attract the attention of senior managers in order for them to form a perception of your abilities. These senior managers may not be directly involved in your work and will therefore pay less attention to your successes. Try to raise your profile subtly with a few of these ideas:

- Circulate useful articles to your project management colleagues or occasionally mention titbits of knowledge you have acquired: 'When I was at the marketing meeting this week Karen told me she was leaving – I thought you might like to know as it could affect the mailing you are planning.'
- Subscribe to magazines relevant to project management or your industry, many of which are free to qualifying applicants. Display them prominently on your desk so your team mates can see you are the sort of person who keeps your knowledge up to date. Starting to raise your profile in your team will have a knock-on effect in other departments, as your colleagues start to mention your name for the right reasons.
- Make sure people know of your achievements. Find opportunities to mention your successes, but remember no one likes a braggart. Slip your self-publicity statement in around another message or story: 'I was out at the weekend with a friend who told me her company has just lost five weeks of client records when their database crashed. It's a good thing I made sure we've got proper back-ups on the project I'm working on, at least that can't happen to the sales database here.'
- Ask to tag along with someone who is attending a meeting with influential people. Either approach it as a work shadowing opportunity or create a reason why it might be relevant to your project.
- Never turn down an opportunity for internal networking: staff briefings, lunches, management question and answer sessions (have an intelligent question to ask).
- Be prepared for your encounters. Plan a 30-second (interesting) summary of your current activity so if you are asked what you're working on you sound intelligent and articulate. You can use this kind of 'elevator' speech at conferences too.
- People like to think they form their own opinions, but in reality they are easily influenced by the opinions of others. If other people think highly of

you, let it be known: 'The editor of XYZ Journal liked the article on risk management I submitted and is going to publish it next month.' It doesn't have to be the opinions of high-flyers or even named individuals either: 'I was chosen to do the analysis on the Project Lambda business case.' But say it in a positive tone: self-deprecation can make even a compliment sound like you were picked because everyone else was working on more exciting and important projects.

- Become indispensable to your programme manager. Stay in the loop about what is happening at programme level and then offer to take the reins when they are away. Do a good job, but hand them back gracefully, with an excellent briefing of what happened. Programme managers will have more confidence in you if you do not become precious about your time in their chair.

- Don't be too modest. Project sponsors want to know that everything is under control, but if you sort out every issue and just reassure them that things are on track, they will just see the smooth progress and not your gargantuan efforts to stay on schedule. In your meetings, explain your difficulty but still add the assurance that you have fixed it: 'We had an issue with our new database servers; there was no space to connect them up in the warehouse. It was a bit frantic, but I finally managed to find the space by getting the team to reorganise what was already in there. They got it all connected in time for the sales team to start testing the new system.' No longer will your hard work be transparent.

- Never talk yourself down, even outside of work. Never use the phrase, 'It's not that interesting.' Someone found your project interesting or they wouldn't have asked you to do it. Even if it is too complicated to explain in detail, you should always remain positive and upbeat about what you do.

You don't actually have to be the best performing project manager in your company to do a good job of promoting yourself. Take the opportunities that are offered, create your own opportunities and build on your self-promotion skills. The trick is making sure that the people who matter think that you are doing a good job regardless of how confident you feel in your own ability.

In order to become recognised for your project management skills make sure you firstly give your senior stakeholders what they want and secondly work on your profile within your organisation.

54 NAVIGATE OFFICE POLITICS

Office politics are a fact of life. Working with others, in any context, means that there will be different personalities on the team or as stakeholders. Juggling the competing priorities of different individuals can be one of the hardest things about project work.

POLITICS IN GOVERNMENT

'For most of my career I have supported various US government agencies,' says Alison Marshall, who now lives and works in the UK. 'In that environment office politics tend to be based on job security and not promotion. The three lines of organisation are government appointed officials, government career staff and contractors. Appointed officials implement policy with a length of service that is limited and undetermined. Career employees enjoy job security and face little repercussion in the workplace for their behaviour. Contractors provide expertise but are vulnerable as they can be removed from their contract at a moment's notice without cause by government staff.

Framed by this work environment, Alison took on the project management of a large electronic data sharing project that would be the first of its kind in the US federal government. The project would impact all career staff at one agency. The stakes were high. 'If the project was successfully implemented within the agency it would be deployed throughout the entire government,' Alison explains. 'A high ranking appointee was driving this project and wanted it completed before his impending retirement. This established an aggressive timeline but it was within the official bounds of his role and he had the authority to provide the resources needed to crash the project schedule.'

While this appointed official wanted the project moved forward quickly, there was another key stakeholder on the project – a career manager who had no interest in whether the project succeeded or failed. 'He enjoyed personal office politics knowing he would not face admonishment,' Alison says. 'He pitted people against each other and fostered the fear of losing their jobs among the contracting staff. His inappropriate interactions would not be tolerated in other workplaces but there was no way to officially stop his behaviour. Members of the project team focused on what they needed to do for self-preservation and not the project.'

Faced with this difficult situation, Alison did what she had to keep the project moving. 'You can't change an individual's personality,' she explains. 'My goal was not to have everyone to play nice but create a structure that promoted civility. When team members worried about job security or that they would be treated badly on a personal level they weren't able to successfully complete their work. I did what I could do to limit the impact of the career manager's inappropriate behaviour on the project.'

With no option of working around this person or having them taken off the team Alison was left with the only solution she could find – personally absorbing as much of the negative office politics as she could in order to shield the team. 'This project required a thick skin and was quite difficult,' she says. 'What I kept in mind was that while I became the focus of this negativity it was important to try not to take it personally. Any person serving as the project manager for this project would have been in the same situation. My determination came from the belief that the project could be successfully delivered and would provide value.'

This approach worked but took a personal toll. 'I was drained by the end of the project but the members of the team felt fulfilled by the scale of their success and that they were not berated in the process,' Alison says. 'Office politics are about people and the motives behind how they interact with each other. My first step on each project is to look at the big picture of office politics surrounding the people officially assigned to each level of organisation. I assess whether their approach to the project simply involves the natural check and balance of their responsibilities or whether non-professional factors are at play. The key is to not confuse role responsibilities or individual work styles with office politics.'

She adds: 'This project was an extreme scenario but it is important to remember that the project manager is always in the cross-fire of office politics. Prepare for office politics by attempting to identify them early. This can prevent you from being caught off-guard when it occurs.'

'Unfortunately, project management has a huge propensity for conflict simply because there are numerous independent relationships and projects often cross organisational and managerial boundaries,' writes Sharon de Mascia in *Project Psychology*.[88] 'The often tight time-constraints of project activities can bring project teams in to conflict with stakeholders who also have to deliver day-to-day business as well.'

Learning that office politics is inevitable is not particularly heartening. Learning how to deal with them is something that takes time and experience, but there are a few things that you can do to make navigating political situations easier.

First, identify if you really are dealing with office politics or if it is a different form of team conflict. Look out for actions that don't seem logical or that don't appear to move the project forward. Listen to the concerns of the team: are they more interested in protecting their own viewpoints, fighting for position or ensuring they get their own way than in what is right for the project? Do they identify with their department or where they came from more than what they are doing now on a cross-functional project team? This could be a sign that the team is divided into silos which will ultimately affect the performance of the team as a whole.

'Once you have identified the sources of the political activity, the next step is to try to understand what the individuals are trying to achieve,' writes Andy Jordan in *Project Pain Reliever*.[89] 'Political games are all about improving the outcome for the players, so as project manager put yourself in the shoes of each of the players and ask yourself what they want.'

When you know what they want, the objective of the politically savvy project manager is to give it to them. This can be straightforward, or it can be difficult; it depends on the people and personalities involved. It can also take a lot of courage to call someone out on behaviour that undermines the project and the team, especially if you personally have something to lose by making the challenge. Sometimes you won't be able to deal with the problem alone, perhaps because you have tried and your efforts have made no difference, or because the player involved is far more senior or influential than you. In that case, you may have to bring in some outside help in the form of your project sponsor, another senior manager or an outside facilitator. Be honest about the challenges facing the project because of politics and if you have any suggestions for how to move the team on from this position, put them forward. Don't be afraid to ask for help from your mentor, manager or project sponsor: that is what they are there for and a project that failed to deliver its objectives because of a team that imploded is never going to look good on your CV.

> Navigating office politics successfully comes with practice. Learn to identify political situations and work to resolve them through creating situations where it is possible for everyone to get what they want in a transparent way without having to undermine the efforts of the project.

55 KNOW WHAT'S A SHOWSTOPPER

A showstopper is something that has stopped or threatens to stop your project today. You will not need to be told how to recognise one; it will be clear from the panicked look on your team's faces and the unspoken question: 'What do we do now?' that hangs silently in the air.

SHOWSTOPPER MANAGEMENT IN ACTION

Eric Spanitz has over 16 years of practical project experience, is a Professor at Lake Forest Graduate School of Management in Chicago on their MBA programme and once trained over 2,500 people on project management techniques for the Canadian military. He is not someone his colleagues would associate with missed milestones. However, even experienced project managers sometimes find it hard to initially realise what could be a showstopper. Eric explains, 'I was managing a project for a medium-sized paper company here in the Chicago area.' The company had around 300 employees and was split across two locations. Eric was working with the IT programmers. 'Now by saying I was "managing" the project,' he continues, 'in hindsight I would say I was cracking the whip, being more of a manager rather than a project manager.'

The project deadlines had been set by senior management without the involvement of the IT department and without a full understanding of the complexity of the project. 'Getting a project done "by Christmas" just had a nice ring to it,' Eric says. 'There was no consideration that I knew of about how realistic this deadline was.' The deadline was particularly tough as between 80 and 90 per cent of the paper company's orders happened during the Christmas season. This put additional pressure on the IT department who had to cope with business-as-usual problem solving and firefighting. Eric, now president of the management consultancy Synergest, says, 'The programmers were all working 14 to 16 hour days, with pretty much everyone coming in on the weekends, myself included, just to even attempt to keep up with the schedule.'

One Friday, Eric came into the office at 8 a.m., late by his normal standards. He had been to see his doctor and was becoming increasingly aware of the ill-health spreading throughout his team. 'I walked by the printer only to see the résumé of our lead programmer printing out at least 100 copies,' he says. 'On my way to my office I passed another programmer, a very proper gentleman in his mid-fifties, with his head on his desk sobbing – no, more like a low wail. Being the emotion-avoider that I was, I ducked into my office to avoid getting pulled into some messy emotional issue. I looked at the clock and realised I had to do something, and at this point really had nothing to lose.'

Eric realised that he had to take drastic action or he would end up losing his entire team. The deadline was completely unachievable and the IT department was falling under a black cloud by trying to do the best they could. 'I loosened my tie, messed up my hair, took a couple of deep breaths and walked down to the conference room where the senior management was having their weekly meeting,' Eric says. 'I walked into the meeting high on adrenaline and just started yelling "you guys are killing us back there – the whole department is sick and getting ready to quit – to hell with your unrealistic deadlines – I'm giving everybody today off and we're not coming in this weekend!" I stormed out of the conference room still seeing red and made a bee-line to the IT department. I quickly whispered loudly to the programmers to grab their stuff and get out – not to come in this weekend, and we'll figure stuff out on Monday. It was like an airplane evacuation as we all ran out the door.'

However, Eric did work that weekend. He used the time to put together a rough project plan for the work, which detailed how it could be completed in a reasonable time frame. His calculations showed that the project could be delivered by the following April. 'I figured even if I was fired, I still would give it to my boss to show why I knew the deadlines were unreasonable,' Eric explains. 'On Monday I went into the Vice President of Operation's office, who was my boss, and handed him the schedule.' Eric didn't know how to read his manager's behaviour. 'Then he said, "Oh good, I thought you were quitting," and looked at the schedule. He said, "You put on quite a performance – we didn't expect that from you." Before I could say anything he said, "I think we need to include you in our planning meetings, so you can tell us, with your inside voice, how realistic our proposed deadlines are."'

Eric came to the realisation that morning that the unrealistic deadline and the decreasing morale of his team was a showstopper. 'If the project manager does not speak up for his or her people, nobody will. Part of a project manager's job is to be an advocate and liaison between the project team and the executives,' he says. Monitoring his team's behaviour allowed him to identify just how low things were getting. 'Extended periods of overtime are counter-productive,' Eric adds. 'At that point, morale is destroyed, people will get sick, people will quit, and productivity plummets. Sitting in front of a computer for longer time periods does not equal more work getting done.'

Since that impromptu meeting with the senior management team, Eric has found that on new projects executives would rarely try to force an unreasonable deadline, as long as he could explain why the deadline was unreasonable.

'Sometimes theatrics are necessary to emphasise a point,' Eric concludes. 'I have never done a similar "performance" again, yet whenever I think back to that situation, I have no doubt that what I did was necessary and appropriate. The senior management team included me as the project manager in all future planning

> sessions involving the IT department. They listened to our discussion about project schedule feasibility, and over the next six months they reworked the pay structure for the programmers to increase pay and to introduce pay for weekend work. I am almost embarrassed to admit that sometimes I wonder if I should have performed sooner...'

You cannot work to resolve or get round the showstopper without first understanding exactly what it is. 'This involves investigation work to determine the dimensions and extent of the problem,' says Richard Murch in his book *Project Management: Best Practices for IT Professionals*.[90] 'Broadly, we define a problem as a deviation from an expected level of performance whose cause is unknown.' Getting to the bottom of the cause is essential, so call a crisis meeting with your team to sit down and work out why the situation has happened and what exactly has gone wrong. There will be a huge temptation to immediately jump in and suggest solutions and a plan of action for what to do next. However, bring your facilitation skills into play and make sure that everybody has the same understanding of the situation and its result before you allow the discussion to move on to analysing options for action.

> Don't be tempted to simply accept the first plan of action. With a little bit of time you and the team may well be able to come up with several alternatives. However, of course in some situations there will only be one alternative: stop the project. In any situation where it is possible to continue the project, stopping it should also be analysed along with the other possible alternatives. This level of consideration is necessary because you and your project team need to be 100 per cent behind the decision. You need to understand why you are implementing this showstopper management plan, and more importantly, be able to explain it to your project board. As the project manager, you may have to take a bullish attitude to move the plan into action as by their very nature, showstoppers don't often give you a lot of time to sit around and deliberate. Some of the most rewarding moments of project management can be in response to this kind of firefighting.

> You will know when a showstopper threatens the success of your project. Keep a calm head, analyse the problem and come up with a robust and effective management plan.

56 LEARN HOW TO FACILITATE

Facilitation is a skill all project managers, all managers for that matter, can benefit from developing. It is a method of participating in a group discussion and working towards the goal of getting the best out of the participants both in terms of their behaviour and also their contribution to the subject up for discussion.

MUSEUM MEETINGS

Naida Kendrick Culshaw was working at a medium-sized science museum in California as their Advertising and Promotions manager during a time of great change for the museum. 'The museum had undergone a large transition from a small gallery space of 20,000 square feet to 100,000 square feet and a sharp increase in staff, who were all moving very quickly and because of that, occasionally out of sync with each other,' she says. 'I approached my manager and said that to do my job efficiently I needed to have a sense of who was doing what and when, as I was responsible for promoting a product – the museum – which I didn't develop or control for the most part.'

Naida proposed holding a cross-departmental meeting as the start of a project team with the aim of bringing together around 15 key personnel to launch new exhibitions in a cohesive way. 'My facilitating skills then came in handy, as I had to manage the individual agendas and concerns of each member of the team,' Naida explains. 'There was a feeling of needing to defend one's turf, and I wanted to bridge the communication gap and engage each individual to see how they could inspire each other to develop solutions to issues that may seem cut-and-dry.'

Naida believes that as a facilitator her role working with this team was to keep the group task-focused, encourage creative thinking, build consensus and keep everyone involved. 'Not everyone approached this as a positive thing, far from it,' she adds. 'But I knew how to manage them – discussing issues offline, helping them see the variety of viewpoints that they may not agree with but are still valid and should be respected. I was the "neutral" person in the room, allowing a free flow of discussion, while keeping things on track.'

This approach paid off, as Naida is quick to explain. 'After we started the process and the members of the team could see the benefits of "opening up" and sharing their ideas and thoughts, we developed a dynamic group that were there to support each other, not to create roadblocks or extra hoops, either on purpose or just by accident.' The first exhibition launch the team worked on together was a great success. 'The team pulled off an amazing exhibition launch, but in the process, I discovered that I had helped to clear long-held misunderstandings by shedding light on the processes of each group and how those impact each other,' Naida says. 'This was the start of a lovely relationship, and in the end, I was asked to be the launch co-ordinator for all exhibitions and films for the museum.'

The cross-functional team may have ended up working together smoothly, but along the route there was still a lot of hard work for Naida. 'For me, the energy spent before, during and after each meeting was listening and watching for nuances, content, body language and other feedback that was being shared by each team member,' she explains. 'By picking up on those signs, I helped facilitate the continued exchange and communication in-between meetings. I had learned from a workshop I attended that in a meeting, or just when speaking with a group, to be aware on two communication levels simultaneously: content – what's being discussed or decided, and process – how the group is functioning.' Naida learnt to blend assertiveness with tact, discipline and humour through the process as well, and became more aware of when to intervene effectively when the meeting veered off the subject.

Naida believes being a good facilitator is partly innate, but partly a learnt skill. 'I really think I sort of "have" this ability,' she says. It has proven a useful skill across her career. After leaving the museum Naida moved to France and became President and VP Programs of WICE, a non-profit cultural and educational association in Paris, where she has used her ability to recruit and motivate a vast team of staff and volunteers. 'It's something I think I honed from childhood. I was the oldest of three, so facilitating the childhood scrapes and arguments was my job, one which I managed with a diplomatic air, according to my mother. My parents always taught us how to look at all points of view, to put ourselves in someone else's shoes, then make comments on a situation.' She points out, however, that facilitation skills can be learnt on a certain level. 'It takes knowledge and talent to read between the lines, to feel a shift in the emotional level in a room and gauge the non-spoken signals that can make or break a session,' she says. 'Those things can be learned, but I feel that those who gravitate towards such a role have a bit of natural talent to build upon.'

Facilitation will not help resolve all situations, according to Naida. 'I think that when you try to facilitate a situation where the parties are just set on their view and will not budge or open up, then facilitating may help them see other points of view – but won't solve the issue.'

'Facilitation is a way of providing leadership without taking the reins,' writes Ingrid Bens in her book *Facilitating with Ease!* 'Most important, you help [group] members define and reach their goals.'[91] Facilitation used to be linked with the set of skills needed to be a good trainer and the 1980s saw facilitation moving out of the classroom with the advent of Total Quality Management (TQM).[92] These days it is well recognised that facilitation skills will help you run project meetings and workshops but 'facilitation' itself is hard to define precisely and you probably do it without being aware of it in many of your meetings already. Try to become more conscious of how you behave in meetings, which will help you identify which areas you need to improve. For example, do you:

- always make sure everyone contributes;
- create an environment where it feels safe to share ideas;
- actively listen to each individual;
- summarise important points and reflect them back to the group;
- ask lots of questions;
- tailor your style depending on the audience and task in hand;
- plan your meetings in advance;
- understand group dynamics, and;
- have experience of how groups resolve issues in a workshop environment?

If you answered 'no' to any of those questions, brushing up your facilitation skills would probably be a good idea. However, you can always bring in an external facilitator, either a consultant, a more experienced project manager or someone from your human resources or training department, to help out for meetings you expect to be particularly tricky to handle. You can offer the same service to other project or operational managers to build up your exposure to these situations.

Judith Kolb and William Rothwell from Pennsylvania State University asked 63 expert facilitators from the International Society for Performance Improvement, an association dedicated to improving productivity and performance in the workplace, about what makes a good facilitator. The responses to their questionnaire showed that the top competencies deemed important for small group facilitators were:

- active listening;
- skilful handling of questions;
- awareness of group dynamics;
- ability to paraphrase contributions from the group, and;
- being able to animate the group and stimulate creativity.[93]

Their study also concluded that there are useful, practical techniques for successful facilitation and a facilitator should have more than one tucked away: 'There are myriad decision making/problem-solving techniques available to help groups manage information and reach decisions,' they write. 'An experienced facilitator should be knowledgeable in a variety of methods so that s/he is not "forcing" a technique that does not fit the situation.'[94]

Tools and techniques will only get you so far. 'Outstanding facilitation is, of course, much more than a smooth presentation style or being adept with flip charts or visual aids,' writes John van Maurik in an article for *Leadership & Organization Development Journal*.[95] 'It is about achieving change, enabling excellence, or empowering groups to achieve results for themselves.' In a project context the results may be a clear understanding of project scope or requirements, an appreciation of what went wrong and what can be done to fix it, the understanding of what other team members do and how they can support each other or a whole host of other objectives. It will be up to you to decide when 'heavy' facilitation is required – with flip charts, post-it notes, warm-up games (ice-breakers) and so on, and when you can get away with facilitating with a lighter touch.

> Some meetings need more than just an agenda and someone to chair. Learning to facilitate can help you get the best out of your time and your team.

57 GET A MENTOR

Having a mentor can be advantageous for your project management career. Mentors offer their skills, experience and, critically, their networks to their protégés or mentees. You can use a mentor to test ideas in a safe environment, ask for advice or just sound off about a particular problem.

MAKING MENTORING WORK

For the first edition of this book, Marina Sampson (now Tumblety) gave her view on having a mentor. When she moved into project management from running a successful customer service team, Marina found an experienced project manager who she felt would be a good mentor. Her mentor at the time worked with her to help her transform her experience of dealing with customers into a new toolkit for dealing with stakeholders and project teams over which she had no line management authority. 'I have learnt so much from my mentor,' Marina said. 'Even though we work in very different styles she will look for the best results for me.' She arranged informal monthly meetings with her mentor for a catch up over coffee and more structured sessions around the technical side of project management if and when they were needed. 'It's important to find someone you can be honest with,' Marina advised then. 'Whatever the issue is we can work together to figure out a resolution.'

Six years on, Marina has progressed to a senior project role and is now managing programmes as well. 'I have taken on a project specialising in distribution deals and website projects,' she says. 'I think I have worked with pretty much all departments and was seeking a new challenge when a secondment into a Lean Six Sigma role came up, offering training, an exam and possible certification so I have grasped that and am learning some more new skills.'

Marina had a mentor up until a few months ago when the individual left to go to another job. 'I have not yet sought another one as I am in a secondment and doing training so I will identify an appropriate mentor once the training and exam are complete,' she says. 'It's something I find really useful and with my last mentor we bounced ideas off one another.'

Marina has also become a mentor herself. 'I have been a mentor for two people, both of whom were in project roles at a lower level,' she explains. 'We took the

> time to set objectives for the sessions and they both progressed into the roles they were seeking. I still meet with them but no longer in a mentor capacity, more for networking now. I would be a mentor again if I was asked and if I felt I could help the person.'

Mentoring can have a very beneficial effect on the protégé's career in terms of improved job satisfaction, more rapid promotion, higher salaries and increased access to organisational key players, according to researcher Kimberly McDowall-Long.[96] However, choosing a mentor should not be rushed into. If your company has a formal mentoring programme you may be allocated a senior manager to support you in the role of mentor. You can still have a mentor even if your company does not have such a scheme, although you will have to seek out and approach them yourself. Either way, hopefully the person chosen as your mentor will play a supportive role in your long-term future.

McDowall-Long identified two groups of characteristics displayed by effective mentors: interpersonal skills and technical expertise. When you begin to think about who you would like as a mentor, consider their competence in both those areas. If you decide to use a formal company scheme ensure you will have the opportunity to swap mentors if the two of you do not click; it can be quite a personal relationship so it is important you get on. Give some thought as well to perhaps having different mentors for different things. The technically brilliant budget whizz may help you out of some awkward financial moments but she may not be the one with whom to discuss sensitive communications issues.

When drawing up your shortlist of potential managers to approach as your mentor consider the following:

- What sort of person do you feel comfortable opening up to?
- How much more senior than you do you want your mentor to be? Perhaps seniority is not as important as their project management experience or their character.
- Would you prefer a male or female mentor? Of what age?
- Do you want someone with an established network who could perhaps help you achieve your career goals?
- Watching them in their interactions with others, do they have the communication skills to be able to give you constructive feedback and set you challenging targets while remaining supportive?

Once you have selected possible candidates, find a convenient time to approach them one by one. Explain to your top choice what you expect of a mentor. How many meetings per month would you want to arrange? Would they be on a regular basis or ad hoc as you need them? How often do you expect to call on them and for what? Once you have outlined your objectives and explained what your prospective mentor is getting into, ask them to consider it and let you know in due course if they are prepared to step into the role. Being asked to be someone's mentor is flattering but it is also a big commitment – informal mentor-protégé relationships can last up to six years.[97] Let your candidate think about it and don't take it personally if they say no. You might have cause to work with them in the future, so don't feel rejected if they turn you down and, of course, maintain a professional relationship. It is unlikely that their decision was based on working with you personally. If they do say no, move on to the second candidate on your list and ask them. Alternatively, ask your top choice politely if they can suggest someone else: they obviously move in the right circles or you would not have selected them, so they may well know of someone outside your personal circle of acquaintance who would be a suitable mentor. And they will probably be pleased that you value their opinion enough to ask.

If the pool of possible candidates within your company has been exhausted, look elsewhere. There are benefits to having a mentor from your own company (increased internal recognition, improved promotion prospects, for example) but there are also advantages to having someone completely independent. There are professional networking groups for all professions, so you should be able to find someone somewhere doing a similar thing to you.

Mentoring can provide you with a great career advantage, access to another set of contacts, and a hotline to technical project management expertise, but you need to choose your mentor carefully in line with your own expectations of how the relationship will develop.

58 DO DOCUMENTATION

Documentation is one of the onerous jobs of a project manager: compiling initiation documents, plans, product descriptions, roles and responsibilities, and the other countless pieces of paperwork (or electronic templates and documents) that form part of your project dossier. However, documentation is essential as it helps turn the nebulous ideas of your project sponsor into a fully fledged, well-understood, tangible project.

Kat Holt, Head of Marketing at the Huntercombe Group, a leading independent provider of health and social care in the UK, realised that her project teams needed a better way to manage their documentation. They chose online project management platform Wrike as a way to store all their project information in one place.

Healthcare is a fast-moving environment and it's essential to provide the best quality services to clients. The marketing team often find themselves dealing with new challenges and conflicting priorities. Old-style paper documentation was no longer an appropriate way to manage project information. The online system allows them to be clear about goals, tasks and upcoming milestones and web access means that important documents are always available, wherever team members happen to be. It has cut out the need to hunt through paper files or multiple software products to find what they need at short notice.

'My team's productivity has increased, as we spend less time trawling through files and spreadsheets to pull documents together,' Kat says. 'All the files and docs we need are conveniently stored in Wrike. It allows us to feel organised in such a fast-paced environment and can manage the emails relating to project tasks for you! All the emails are streamlined into one user interface that saves lots of time on going from one email to the next and copying assignments.'

Project documentation – in the widest sense – serves three functions: clarification, concentration and confirmation.

- **Clarification:** the act of writing things down allows you to clarify vague ideas. It guarantees that everyone involved in the project has a clear understanding of the aims and objectives, scope, plan and assumptions. The act of working with key stakeholders to create these documents or enter objectives and so on into an online project management system facilitates discussion around what everyone is expecting from the project. Any vagaries or multiple interpretations can be cleared up at this point.

- **Concentration:** a roles and responsibilities statement sets out exactly what is expected of each member of the project team, at every level. It ensures people know what they're signing up for and is especially useful as a starting point for discussion with the project sponsor about their role, as senior managers are often not clear about what sponsoring a project entails. Documentation encourages people to concentrate in another way too – the review and sign-off process ensures you have agreement from all key decision makers. Whether this is a physical signature on a document or approval via an email, the act of sign off is the same: it's tantamount to an informal, internal contract. While managers do approve project documents without reading them, asking them to put their sign off in writing can increase the pressure to read the documentation. It will never guarantee complete buy-in, but it is a step in the right direction. To check if her documentation was read, one project manager issued a draft project initiation document that listed one of the responsibilities of her sponsor as sending her chocolate on a weekly basis. The sponsor noticed it, but none of the other document reviewers mentioned it at all.

- **Confirmation:** while a project only exists in people's heads it will struggle to be taken seriously. Sarah Blackburn, in her paper on project networks, explains that 'the project also has to create itself and maintain itself for the time needed to effect delivery… [it] craves embodiment: a code name, a project room, a logo, t-shirts – these are peripherals – the core is the project documentation.'[98] Documentation, whether stored electronically or physically on paper, makes the project 'real'. It moves the project from the realm of a good idea into a realm where a professional project manager can turn it into a tangible activity and delivery.

For your project, consider what documents you will need to produce. As the minimum you will need those covered in Table 58.1. Your organisation might expect you to produce a quality plan or other documentation too. It might seem like a lot of paperwork but incomplete or missing documentation has been identified as an early warning sign for projects that are likely to struggle or fail[99] so it really is worth taking the time to complete everything.

Once a document is written it should be circulated for review by anyone who will be impacted by the content. Online tools with workflow capability are one way to do this. Give your reviewers a deadline by which to send you their comments and add that if you hear nothing you will assume that they have no comments to make. Include the feedback you have received and if it has changed the document substantially, reissue it for review. An example of a substantial change would be if a reviewer from marketing added in additional work for the sales team to do. Changes to punctuation or layout would not require the document to be reissued for a further review. Reviewers should comment in detail on parts that impact their own team, but their comments about the involvement of other teams should always be discussed with someone from that team. When all the substantial changes have been validated by the right people, the document is ready to be finalised. In your covering email or a note attached to the document in an

Table 58.1 Standard project documents

Document	Function
Project initiation document (PID)	The PID comes in many forms and is the first project document produced after approval is given for the project to start. It covers project scope, objectives, the principles and methodology under which the work will be carried out, high-level budget and may incorporate high-level plan details and the initial risk and issue log.
Plan	A plan can be made up of sub-documents, e.g. a detailed description of all the deliverables. As a minimum, a project plan requires a schedule which is a list of the tasks to be done, plus the dates they will start and finish and who is to do them.
Roles and responsibilities statement	This explains the roles within the project (sponsor, project manager, business owner, IT team leader and so on) along with the responsibilities that role carries. It should be as specific as possible. Electronic document management systems may also use roles and responsibilities to grant access to documents and project information.
Initial risk and issue log	The risk and issue log is a living document that will be added to as you go through the project and continue to manage existing risks and issues and add new ones. At the beginning of the project, this document sets out what it is you know already about the risks and issues associated with the project. Storing this information in an electronic project management system makes it easier to update and ensures everyone has access to the latest version.
Requirements document	This details exactly what any new process, product or system should do and is put together based on the customer/business requirements.
Post-project documentation	At the end of the project, you will need to get approval for the project to be closed down. This is normally in the form of a post-project document, which follows a post-project review meeting. A shorter document, a project close-down document, can be used instead for formal authority to stop the project, especially if the project did not reach its planned end.

online repository make a note of any minor changes you have made that the approver has not yet seen.

Whether you have a hard copy of the document or an electronic one, it is useful to keep track of who has approved it. If your online project management system will not do this for you, add a simple approval area to the document as shown in Table 58.2.

Table 58.2 Example of an approval area on a typical project document

The approvers below authorise the work detailed in this document to be carried out

Name	Role	Signature	Date
Meera Rantasha	Project Sponsor, Director of Operations		
Simon Wilkinson	Project Manager		
Natasha Culshaw	Head of Sales		

On a hard copy of the document, approvers can physically sign their names. On an electronic copy, you can annotate the signature box with a note saying 'approved by email' and the appropriate date. Embed copies of the relevant emails into the document or add them to your electronic filing system to give you an audit trail of approvals. See Chapter 19 for more on version control for documents.

> By producing documentation, whether electronic or in paper format, you are giving the project shape and structure, helping to generate buy-in and ownership and also ironing out any vagaries early on.

59 DON'T BE AFRAID TO SUGGEST THEY PULL THE PLUG

The average large company, running around 150 projects at any one time, loses £13 million a year by not stopping projects that are failing.[100] Suggesting your project is scrapped can be a difficult message to give your sponsor and stakeholders. However, the project manager's role is to partly direct the work and partly to provide an objective position on how the work is done and if that means suggesting stopping everything and starting again, or even not starting again, then that is part of your role too.

SECURING A GOOD RESULT

A junior manager working in the security industry was called in to help with the communications on a particularly sensitive project. A new piece of legislation meant all security guards had to be licensed, a process which cost around £200 per employee. The project team had been working on applying for licences within the deadline and the payment process to pass the cost on to the individual guard. The manager recognised the human implications of passing a £200 bill on to staff who, in the main, were receiving minimum wage. 'The impact hadn't been thought through,' he says. When he fully understood the project's key stakeholders he immediately approached the right people and recommended the initiative was stopped. The alternative proposal was for each of the security company's clients to meet the bill for licensing, instead of passing the charge on to the employee. The project was restarted with this as the end goal, saving the company an embarrassing situation with unions and staff.

If you continue to work on a project that you know is doomed you risk being personally tarnished with the project's failure and being criticised for not making your superiors aware of the situation. Even if you do make senior executives aware of the inevitability of the situation, your sponsor may well insist on the work being completed even once you have presented the most compelling of arguments for why that is not a practical conclusion. The important point for you is that you will have done your bit in highlighting the situation and will, of course, maintain your risk and issue log regularly to make certain that your concerns are noted and action is being taken to mitigate the risk of failure.

Keeping note of those risks and issues during a project which is on the brink of failure is particularly important. If your sponsor will not listen to your advice to cut the project's losses and get out now, you can at least use the risk and issue log to explain the consequences of completion. Sometimes that alone can have a sobering effect. There is nearly always another way to achieve similar objectives, so if you can think of another solution, point it out.

There are sensitivities to take into account of pointing out that a project is never going to complete successfully. In some cultures, regardless of what you believe, having that discussion with your project sponsor would be unthinkable. 'Failure, operationalised as a cancelled project or a bankrupt company, is not as grave in Silicon Valley as it would be in Taipei, where success and failure is bound up with mianzi, or "face", a concept that is culturally specific,' write three anthropologists in their paper 'Trusting strangers'.[101]

> Do not be afraid to challenge senior people. Remember that not all projects are started from a basis of a well-thought-through and competent idea. If the project is going to be a failure and you know it, explain why it should be stopped.

60 ARCHIVE EFFECTIVELY

It is the end of your project: the sponsor thinks you did an excellent job and the people who have to live with the change are happy. But it's not quite over. All those documents, minutes, decisions, test notes and emails that you have accumulated over the length of the project have to go somewhere and it would not be right to dump them in the recycling bin. This is where archiving comes in. At some point in the future someone is bound to ask you why a particular task was or was not done. Archiving will mean you can get hold of the information easily to answer their question, but the files won't be cluttering up your desk or network drive while you are trying to work on something else. Having the files available will also help other project managers who find themselves doing a similar project in the future.

ARCHIVING FOR ARCHIVES' SAKE

John, a journalist, was updating the archives of the regional south of England weekly newspaper where he was working when he came across a file called 'Bubbles'. 'I wasn't sure why the paper would have a whole archive folder on the subject of bubbles – we used the archives as a reference for future new stories,' he says. When John looked, there was one clipping in the folder. The story was about a family reunion with the headline 'Daughter bubbles with joy'.

As a rule of thumb, keep your files easily accessible for one year. Then archive them. Follow any governance policies or regulatory guidelines for storing project records – if you don't know what may apply to your project, talk to your senior executives or the company's legal team.

Electronic records stored in a document repository or as part of online project management software can get very messy if you don't have some guidelines about how to store them long term. Take some advice from your Project Management Office or other project managers if you need help working out how best to make your records accessible for the future.

Remember that your project records also include the information that is in people's heads – recording lessons learnt sessions can be effective. The very fact of having a video camera or microphone in the room can make a meeting more effective as people subconsciously react to being recorded. You can also create webinar versions of key presentations and record question and answer sessions – all materials that could be useful in the future to other project teams or the business as usual team responsible for the ongoing management of the product your project has created.

If you are using archive boxes to move paper records off-site and into storage, make your box and folder labels transparent and logical. As your project files may be used by someone else in the future don't be cryptic – it might even be you in a few years who requires access to information and what was crystal clear then may have faded in your memory with time.

A project isn't over when the post-project review is signed. Archive your project data to be sure the full history is available if it is ever needed again.

FURTHER READING FOR THIS SECTION

Asher, D. (2007) *Who Gets Promoted, Who Doesn't and Why*. Ten Speed Press, Berkley.

Franklin, M. and Tuttle, S. (2008) *Team Management Skills for Project and Programme Managers*. TSO, London.

Keegan, A. E. and Den-Hartog, D. (2004) 'Transformational leadership in a project-based environment: a comparative study of the leadership styles of project managers and line managers'. *International Journal of Project Management*, 22, 609–617

Kirk, P. and Broussine, M. (2000) 'The politics of facilitation'. *Journal of Workplace Learning: Employee Counselling Today*, 12 (1)k, 13–22.

Laborde, G. Z. (2001) *Influencing with Integrity: Management Skills for Communication and Negotiation*. Crown House Publishing, Carmarthen.

Mersino, A. (2007) *Emotional Intelligence for Project Managers*. Amacom, New York.

Pinskey, R. (1997) *101 Ways to Promote Yourself: Tricks Of The Trade For Taking Charge Of Your Own Success*. Avon, New York.

Robbins, H. and Finley, M. (2004) *The Accidental Leader*. Jossey-Bass, San Francisco.

Turner, R. J. (2006) 'Programme and portfolio management: connecting projects to corporate strategy'. *Project Manager Today*, January 2006, 13–16.

JOIN THE CONVERSATION

Got a case study to share? Want some feedback on your own project experiences? One of the aims of writing this new edition was to highlight some of the great things that project managers and project teams are doing to make their projects better. I'd love to hear about the ways that you are implementing some of the ideas in this book (or your own good practices) in your organisation. You can find me online at www.GirlsGuideToPM.com or on Facebook at www.facebook.com/Elizabeth.Harrin. Please stop by and say hello.

SHARE THESE IDEAS

I think that project managers have a wealth of knowledge to share but not much time to do so. If you think that the ideas and practices in this book could help your colleagues, you can help spread the word. Please consider doing something to help these ideas reach a wider audience. Here are some suggestions:

- Write a review of this book on Amazon or your favourite online bookstore.
- Mention this book on Twitter, LinkedIn, your blog or your favourite social media site.
- Leave this book prominently on your desk for your colleagues in your PMO to flick through.
- Talk about the ideas in this book at a 'lunch and learn' session, and share the concepts with your colleagues.
- Lead by example! Demonstrate your new skills and knowledge and let others learn from you.

Thank you for your support.

APPENDIX 1
RISK LOG

Impact categories

Minor: impact of less than £40,000 or no reputational damage; Moderate: impact of between £40,000 and £200,000 or possible small amount of reputational damage; Significant: impact of between £200,000 and £1 million or moderate amount of reputational damage; Severe: impact of over £1 million or destructive reputational damage

Likelihood categories

Remote: less than 1 in 1,000 chance of occurring – not foreseeable within five years; Unlikely: less than 1 in 100 – could happen within five years; Possible: less than 1 in 10 – could happen within a year; Probable: less than 1 in 5 – imminent

Risk ID	Title	Date Raised	Owner	Impact	Likelihood	Status	Notes and Actions

APPENDIX 2
ISSUE LOG

Priorities
High, Medium, Low

Issue ID	Title	Date Raised	Owner	Priority	Status	Notes and Actions

APPENDIX 3
CHANGE LOG

Priorities
Critical, Important, Cosmetic, Optional

Outcome categories
Reject, Accept, Pend/Postpone

Change ID	Description	Date Raised	Owner	Related Documents	Priority	Date Impact Assessment Completed	Outcome

APPENDIX 4
FOREWORD TO THE FIRST EDITION

In this book Lonnie Pacelli is quoted as saying 'Surprises are for birthdays...' It was in fact a few days after my birthday that Elizabeth Harrin approached me to write this foreword, a very pleasant surprise!

In the media we frequently read or hear about project failures, which consequently adversely affect the reputation of all project managers. The success stories rarely hit the headlines. As Chair of the British Computer Society's Project Management Specialist Group (BCS PROMS-G) I am, **inter alia**, responsible for promoting professionalism in our specialist sector of the IT industry. By providing, through our countrywide events, timely and relevant information on industry developments and by sharing lessons learnt, PROMS-G promotes continuing professional development. Our aim is that project managers, and therefore their projects, will be increasingly successful and hit the headlines for the right reasons! Elizabeth is one of our 5,000-plus valued members and an occasional speaker.

A key skill required of all project managers is to identify potential risks and to remove or mitigate their effect before they become issues. Whilst we all appreciate pleasant surprises, it is the unpleasant ones that have the most adverse effects on a project. Whether you are an experienced project manager or not it is highly probable that you will come across both types of surprise.

It is, however, impossible for project managers to foresee all situations that may arise. While we should all attempt to continually develop our professionalism and to keep abreast of developments in our own industry or particular area of expertise this may not always be possible due to the large amount of change that occurs. It is therefore imperative that project managers are able to focus on these changes and assess their impact rather than spend their precious time resolving underlying project management issues such as budgets, processes and so on.

This book aims to assist with getting the latter right. It is a valuable reference point for ensuring that a project has the underlying essential processes and authorities in place and that they are working as intended. Some of the pitfalls that await the unwary or unskilled are identified and guidance is provided on how to avoid them. In following these recommendations, and not spending time resolving basic issues, a project manager's time will increase allowing him or her to focus instead on the more critical risks and issues.

It is no surprise to me that Elizabeth has written a book that is very easy to read; you can dip in and out of it as required. Each part is self-contained and will provide that nugget of information you have been looking for. Elizabeth has collected the issues, anecdotes and success stories not of entire projects but of the elements within them. I am pleased that so many project managers were willing to share their experiences because it is only by sharing and learning from these experiences that we can all continually develop and enable our professionalism to grow. All project managers, whether working in IT or in other industries, will easily identify with the lessons learnt. If you find something works for you then please pass it on. By the way, PROMS-G is always looking for speakers for our events!

The phrase 'Surprises are for birthdays...' is one of the mantras that should guide us in all aspects of project management. As a professional project manager and Chair of PROMS-G perhaps I should perhaps have anticipated the pleasant surprise of being asked to write this foreword. On the whole though I would rather focus on avoiding the unpleasant surprises and leave the pleasant surprises just as they are. Elizabeth's book helps to do just that.

Ruth Pullen
Chair, Project Management Specialist Group
www.proms-g.bcs.org
British Computer Society
March 2006

END NOTES/REFERENCES

1. Sauer, C. and Cuthbertson, C. (2004) 'The state of IT project management in the UK 2002–2003'. *Computer Weekly*. www.computerweeklyms.com/pmsurveyresults/surveyresults.pdf (5 March 2006).

2. For more on this project, see GAO-04-611 (2004) *Nuclear Waste: Absence of Key Management Reforms on Hanford's Cleanup Project Adds to Challenges of Achieving Cost and Schedule Goals*. US Government Accountability Office, Washington DC.

3. GAO-05-123 (2005) *DOE's Management of Major Projects*. US Government Accountability Office, Washington DC.

4. GAO-09-913 (2009) *Uncertainties and Questions about Costs and Risks Persist with DOE's Tank Waste Cleanup Strategy at Hanford*. US Government Accountability Office, Washington DC.

5. Doss, G. M. (2005) *IS Project Management Handbook*. Aspen, New York, p. 61.

6. Shim, J. K and Siegel, J. G. (2005) *Budgeting Basics and Beyond*. Wiley, New York. Reprinted with permission of John Wiley & Sons, Inc.

7. Cavanagh, M. (2012) *Second Order Project Management*. Gower, Farnham, p. 81.

8. Source of image: State of Western Australia

9. Keil, M., Mann, J. and Rai, A. (2000) 'Why software projects escalate: An empirical analysis and test of four theoretical models'. *MIS Quarterly*, 19 (4), 421–447, cited in Zhang, G. P., Keil, M., Rai, A. and Mann, J. (2003) 'Predicting information technology project escalation: A neural network approach'. *European Journal of Operational Research*, 146, 115–129.

10. Flyvberg, B., Holm, M. K. and Buhl, S. L. (2002) 'Understanding costs in public works projects: error or lie?' *Journal of American Planning Association*, 68, 279–295, cited in Eden, C., Ackermann, F. and Williams, T. (2005) 'The amoebic growth of project costs'. *Project Management Journal*, 36 (1), 15–27.

11. Andersen, E. S. (2010) 'Are we getting any better? Comparing project management in the years 2000 and 2008'. *Project Management Journal*, 41 (4), 4–16.

12. Fleming, Q. W. and Koppelman, J. M. (2005) *Earned Value Project Management*, Third Edition. PMI, Newtown Square, p. 114.

13. Schulte, R. (2002) *Modern Cost Management*. Welcom White Paper. www.welcom.com/content.cfm?page=143 (21 April 2006).

14. Wright, J. N. (1997) 'Time and budget: the twin imperatives of a project sponsor'. *International Journal of Project Management*, 15 (3), 184.

15. Heldman, K. (2003) *Project Management JumpStart*. Sybex, San Francisco, p. 206.

16. Ibid., p. 207.

17. Williams, T. C. (2011) *Rescue the Problem Project*. Amacom, New York, p. 8.

18. Finch, C. (2007) *All Your Money Won't Another Minute Buy: Valuing Time as a Business Resource*. Journyx, Austin, p. 7.

19. Research commissioned by Microsoft and Corporate Project Solutions, cited in Lane, K. (2004) 'Worrying flaws in PM practice'. *Project Manager Today*, April 2004, 4.

20. www.changefirst.com/uploads/documents/For_every_pound_you_spend_on_change_management,_your_company_gets_seven_back_-_ROI_survey_results_-_Changefirst_Article.pdf (2 January 2013).

21. Sauer, C. and Cuthbertson, C. (2004) 'The State of IT Project Management in the UK 2002–2003'. *Computer Weekly*. www.computerweeklyms.com/pmsurveyresults/surveyresults.pdf (5 March 2006).

22. www.youtube.com/watch?v=uqs1YXfdtGE

23. www.metoffice.gov.uk/

24. This six-step approach has been adapted from the model presented in Everett, C. (2005) 'How to ensure pilot projects are successful'. *Computing*, 3 November 2005, 48–52.

25. Office of Government Commerce (2008) *Portfolio Programme and Project Offices*. The Stationery Office, Norwich, p. 5.

26. Durbin, P. and Doerscher, T. (2010) *Taming Change with Portfolio Management*. Green Leaf Press, Austin, p. 41.

27. Wood, M. (2012) 'The team is frustrated with rework based on changing requirements'. In Garrett, D. (2012) *Project Pain Reliever*. J Ross Publishing, Fort Lauderdale, pp. 216–219.

28. Furman, J. (2011) *The Project Management Answer Book*. Management Concepts, Vienna, p. 124.

29. There isn't space to discuss this in more detail here, but you can see *Customer-Centric Project Management* by Elizabeth Harrin and Phil Peplow for a model to use for engaging stakeholders in the discussion about quality and value.

30. Crawford, L. (2005) 'Senior management perceptions of project management competence'. *International Journal of Project Management*, 23, 7–16.

31. Cooke-Davies, T. (2002) 'The "real" success factors on projects'. *International Journal of Project Management,* 20, 188.

32. White, T. (2004) *What Business Really Wants from IT: A Collaborative Guide for Business Directors and CIOs*. Butterworth-Heinemann, Oxford, p. 111.

33. Thomas, J. and Mullaly, M. (2008) *Researching the Value of Project Management*. PMI, Newtown Square, p. 350.

34. Le Guin, U. K. (1998) *Steering the Craft*. Eighth Mountain Press, Portland, p. 33.

35. Raynes, M. (2002) 'Document management: is the time now right?' *Work Study*, 51 (6), 303–308.

36. You can read more about this case study in *Customer-Centric Project Management* by Elizabeth Harrin and Phil Peplow (Gower, 2012).

37. The author acknowledges Martin Schindler and Martin J. Eppler for the term 'project amnesia' in the paper Schindler, M. and Eppler, M. J. (2003) 'Harvesting project knowledge: a review of project learning methods and success factors'. *International Journal of Project Management*, 21.

38. HC1159-11 (2004) *Ministry of Defence, Major Projects Report 2004*, Project Summary Sheets. Report by Comptroller and Auditor General, The Stationery Office Ltd, London.

39. HC1159-1 (2004) *Ministry of Defence, Major Projects Report 2004*. Report by Comptroller and Auditor General, The Stationery Office Ltd, London.

40. Elkington, P. and Smallman, C. (2002) 'Managing project risks: a case study from the utilities sector'. *International Journal of Project Management*, 20, 56–57.

41. Risks do not have to be limited to negative outcomes, although this is the traditional way of defining project risk. For an analysis of how to extend the risk management process to include the management of positive risk, see Hillson, D. (2002) 'Extending the risk process to manage opportunities'. *International Journal of Project Management*, 20, 235–240.

42. For more information about risk logs and how risks can be logged in practice, see Patterson, F. D. and Neailey, K. (2002) 'A risk log database system to aid the management of project risk'. *International Journal of Project Management*, 20, 365–374. This paper provides an overview of a risk database used in an automotive company and is a useful study into how risk logs can be made to work.

43. Cooke-Davies, T. (2002) 'The "real" success factors on projects'. *International Journal of Project Management*, 2,0, 186.

44. Elkington, P. and Smallman, C. (2002) 'Managing project risks: a case study from the utilities sector'. *International Journal of Project Management*, 20, 55.

45. Baccarini, D., Salm, G. and Love, P. E. D. (2004) 'Management of risks in information technology projects'. *Industrial Management & Data Systems* 104 (4), 291.

46. See Hubbard, A. 'Football's coming home' in Project, October 2012, for more on this case study.

47. See, for example, Khalfan, A. (2003) 'A case analysis of business process outsourcing project failure and implementation problems in a large organization of a developing nation'. *Business Process Management Journal*, 9 (6) 745–759; Dvir, D., Raz T. and Shenhar, A. J. (2003) 'An empirical analysis of the relationship between project planning and project success'. *International Journal of Project Management*, 21, 89–95; Turner, J. R. (2004) 'Five necessary conditions for project success'. *International Journal of Project Management*, 22, 349–350.

48. I am indebted to Neville Turbit for explaining this exercise to me.

49. Goodman, J. and Truss, C. (2004) 'The medium and the message: communication effectively during a major change initiative'. *Journal of Change Management*, Taylor & Francis Ltd, 4 (3), 225–226 (www.tandf.co.uk/journals).

50. Obeng, E. (2003) *Perfect Projects*. Pentacle Works, Beaconsfield, p. 126.

51. Trust in a workplace environment is discussed in an interesting study by English-Lueck, J. A., Darrah, C. N. and Saveri, A. (2002) 'Trusting strangers: work relationships in four high-tech communities'. *Information, Communication & Society* 5 (1), 90–108.

52. Thanks to Phil Peplow for providing input to this section and concept.

53. See, for example, Takeya, M. (2009) 'Make project sponsors write their own requirements', in Davis, B. (ed) *97 Things Every Project Manager Should Know*. O'Reilly, Sebastopol.

54. White, D. and Fortune, J. (2002) Current practice in project management – an empirical study. *International Journal of Project Management*, 20.

55. Trompenaars, F. and Woolliams, P. (2003) 'A new framework for managing change across cultures'. *Journal of Change Management*, 3 (4), 361–375.

56. The author is grateful to Barry Shore and Benjamin J. Cross for posing this question in the paper: Shore, B. and Cross, B. J. (2005) 'Exploring the role of national culture in the management of large-scale international science projects'. *International Journal of Project Management*, 23, 55–64.

57. *Distributed Agile Teams: Achieving the Benefits*. Research report by ProjectsAtWork.com. www.projectsatwork.com/content/White-Papers/272852.cfm (6 January 2013).

58. GAO-01-459 (2001) *Computer-based Patient Records: Better Planning and Oversight by VA, DOD and IHS would Enhance Health Data Sharing*. US Government Accountability Office, Washington DC.

59. GAO-02-703 (2002) *Veterans Affairs: Sustained Management Attention is Key to Achieving Information Technology Results*. US Government Accountability Office, Washington DC.

60. GAO-04-687 (2004) *Computer-based Patient Records: VA and DOD Efforts to Exchange Health Data Could Benefit from Improved Planning and Project Management*. US Government Accountability Office, Washington DC.

61. GAO-07-755T (2005) *Systematic Data Sharing Would Help Expedite Service Members Transition to VA Services*. US Government Accountability Office, Washington DC.

62. White, D. and Fortune, J. (2002) 'Current practice in project management – an empirical study'. *International Journal of Project Management*, 20.

63. You can read more about this case study in *Customer-Centric Project Management* by Elizabeth Harrin and Phil Peplow (Gower, 2012).

64. Hacker, M. (2000) 'The impact of top performers on project teams'. *Team Performance Management*, 6 (5/6), 88.

65. If you are lucky enough to get to choose your own team, Meredith Belbin's research into team roles might be interesting: www.belbin.com

66. Obeng, E. (2003) *Perfect Projects*. Pentacle Works, Beaconsfield, p. 107.

67. Boddy, D. and Paton, R. (2004) 'Responding to competing narratives: lessons for project managers', *International Journal of Project Management*, 22, 231.

68. www.sparksnetwork.org

69. Taylor, P. (201) *Leading Successful PMOs*. Gower, Farnham.

70. Juli, T. (2011) *Leadership Principles for Project Success*. CRC Press, Boca Raton, p. 172.

71. Bloch, S. and Whiteley, P. (2009) *How to Manage in a Flat World: 10 Strategies to get connected to your team wherever they are*. Pearson, Upper Saddle River. p. 132.

72. Center for Business Practices (2003) *Project Management: The State of the Industry*. A summary of this research report is available online at www.pmsolutions.com/articles/pdfs/general/industry_news.pdf (3 March 2006).

73. This case study is based on a presentation given by Henrik Kniberg at Øredev software development conference in Mälmo, Sweden, November 2012.

74. Lissak, R. and Bailey, G. (2002) *A Thousand Tribes: How Technology Unites People in Great Companies.* Wiley, New York.

75. White, D. and Fortune, J. (2002) 'Current practice in project management – an empirical study'. *International Journal of Project Management,* 20.

76. Deckro, R. F. and Hebert, J. E. (2002) 'Modelling diminishing returns in project resource planning'. *Computers & Industrial Engineering,* 44, 20. This paper presents a series of models for identifying the point of diminishing return.

77. Shrub sets out his project planning model for project segmentation in his paper: Shrub, A. (1997) 'Project segmentation – a tool for project management'. *International Journal of Project Management,* 15 (1), 15–19.

78. Reiss, G. (1995) *Project Management Demystified: Today's Tools and Techniques.* Spon Press, London, p. 51.

79. Herroelen, W. and Leus, R. (2005) 'Project scheduling under uncertainty: Survey and research potentials'. *European Journal of Operational Research,* 165, 289.

80. Herroelen, W. and Leus, R. (2004) 'The construction of stable project baseline schedules', *European Journal of Operational Research,* 156, 550.

81. See for example, Herroelen, and Leus' 2004 article, which sets out a mathematical programming model for designing a baseline project schedule.

82. CERFDOE Final Report – 071204 (2004) *Independent Research Assessment of Project Management Factors Affecting Department of Energy Project Success.* Civil Engineering Research Foundation, United States.

83. 3M survey (1998) cited in Simon, P. and Murray-Webster, R. (2005) 'Efficient and effective meetings: essential but elusive?' *Project Manager Today,* June 2005, 16.

84. Garcia, A. C. B., Kunz, J. and Fischer, M. (2005) 'Voting on the agenda: the key to social efficient meetings'. *International Journal of Project Management,* 23, 20.

85. Rowe, S. (2007) *Project Management for Small Projects.* Management Concepts, Vienna, p. 6.

86. Mark, G., Gonzalez, V. M. and Harris, J. (2005) 'No task left behind? Examining the nature of fragmented work', in CHI 2005 PAPERS: *Take a Number, Stand in Line (Interruptions & Attention 1)* April 2–7, Portland, Oregon, USA.

87. Crawford, L. (2005) 'Senior management perceptions of project management competence'. *International Journal of Project Management,* 23, 7–16.

88. De Mascia, S. (2012) *Project Psychology: Using Psychological Models and Techniques to Create a Successful Project.* Gower, Farnham, p. 8.

89. Jordan, A. (2012) 'The office politics are killing me', in Garrett, D. (ed) *Project Pain Reliever*. J. Ross, Fort Lauderdale, p. 125.

90. Murch, R. (2001) *Project Management: Best Practice for IT Professionals*. Prentice Hall, Upper Saddle River.

91. Bens, I. (2005) *Facilitating with Ease!* Jossey-Bass, San Francisco.

92. Nelson, T. and McFadzean, E. (1998) 'Facilitating problem-solving groups: facilitator competences'. *Leadership & Organization Development Journal*, 19 (2), 73.

93. Kolb, J. A. and Rothwell, W. J. (2002) 'Competencies of small group facilitators: what practitioners view as important'. *Journal of European Industrial Training*, 26 (2/3/4) k, 201.

94. Ibid., p. 202.

95. van Maurik, J. (1994) 'Facilitating excellence: styles and processes of facilitation'. *Leadership and Organization Development Journal*, 15 (8), 34.

96. McDowall-Long, K. (2004) 'Mentoring relationships: implications for practitioners and suggestions for future research'. *Human Resources Development International*, 7 (4), 519–534.

97. Ragins, B. R. and Cotton, J. L. (1999) 'Mentor functions and outcomes: a comparison of men and women in formal and informal mentoring'. *Journal of Applied Psychology*, 84 (4)k, 529–549 cited in above.

98. Blackburn, S. (2002) 'The project manager and the project-network'. *International Journal of Project Management*, 20, 202.

99. Klakegg, O. J. et al. (2010) *Early Warning Signs in Complex Projects*. PMI, Newtown Square.

100. Research commissioned by Microsoft, cited in *Computer Weekly*, 17 May 2004.

101. English-Lueck, J. A., Darrah, C. N. and Saveri, A. (2002) 'Trusting strangers: work relationships in four high-tech communities'. *Information, Communication & Society* 5 (1), 95.

INDEX

A Thousand Tribes (Lissak/Bailey), 142
actual cost of work performed (ACWP), 12
Agile projects, 64, 101, 141, 150, 158, 174
All Your Money Won't Another Minute Buy (Finch), 36
Amazon, 202
ambiguity, 63–65
American Hearing Loss Association, 149
Apollo 11 mission, 146
archiving, 200–201
Areas of Outstanding Natural Beauty (AONB), 21–22
Assessment Phase, 74
assumptions, 86–88
Australian Computer Society Project Management group, 86
Australian National Competency Standards, 178
automated calendar events, 165

Baccarini, D., 81
Bailey, George, 142
baselines, 61–62
Battersea Power Station, 147
benefits
    financial, 41
    flexibility, 41
    internal, 41
    management, 61
    mandatory, 40–41
    productivity, 41
    quality of service, 41
    risk management, 42
Bens, Ingrid, 188

Blackburn, Sarah, 195
Bloch, Susan, 136
brainstorming, 8, 60, 64, 89
budgets
    change management, 38–39
    contingency fund, 15–16, 18–20
    definition, 8–9
    no-budget projects, 33–34
    overspending, 12–13
    project, 4–6, 30
    project management, 12, 27
    real time, 11–12
    realistic, 3
    responsibility, 21–25
    tolerance, 14–17
    tracking projects, 11, 26
business analysis, 65

Campbell, Sharon, 149–150
Cavanagh, Michael, 9
Centre for Complexity and Change, Open University, 112
change, 93–96, 101, 207–208
change control/management, 52–54
Changefirst, 38–39
clip art pictures, 126
commitment, 105, 107
communication, 93–96, 122
Computers & Industrial Engineering (Deckro/Hebert), 148
contractor costs, 24
Cooke-Davies, Terry, 59
costs
    calculating true, 7
    management/deliverable, 8–9
    of projects, 4–5, 8, 25

Craig, Paul, 45–46
Crawford, Lynn, 57, 178–179
critical path analysis, 150–151, 152–154
Culshaw, Naida Kendrick, 187–188

'daily cocktail party', 158
Deckro, Richard, 148
Department of Defense (DOD), 111
Department of Industrial Engineering, Tel Aviv University, 151
Department of Veteran Affairs (VA), 111
Dijkhuizen, Leo, 51–52
Doerscher, T., 50
Duarte, Nancy, 127
Dupleix, Marie-Hélène, 26
Durbin, P., 50
Duthie, Ian, 55, 58

earned value analysis (EVA), 13
Earned Value Management, 30
Earned Value Project Management (Fleming/Koppelman), 19–20
Elkington, P., 75
ESI International, 130
estimating
    estimate at completion (EAC), 12–13
    estimate to complete (ETC), 11–13
    note on, 5
expenditure, 6, 8–9, 13, 19, 23–24, 26–27, 30, 61
extreme-ultraviolet imaging spectrometer, 77, 83

Facebook, 177–178
Facilitating with Ease! (Bens), 188
facilitation, 187–190
feedback, 47, 113–114
Financial Services Authority (FSA), 105
Finch, Curt, 36
Fish, Michael, 45
Fleming, Q. W., 19–20
Furman, Jeff, 56

Garcia, Ana, 159
George, Gina, 100–101, 165–166
Gideon Reeling, 146–147
glass recycling, 153–154
golden triangle, 43, 56
Goodman, Joanna, 94
Google Docs, 5
Government Accountability Office (GAO), 4, 111–112
graphic recording, 127

Hacker, Marla, 114–115
Hammond, Val, 21–23
Hanford nuclear processing plant, 3–4
Harrin, Elizabeth, 209–210
Harvey, Gordon, 32–33
health and safety, 57, 146–147, 164
healthcare, 194
Hebert, John, 148
Heldman, Kim, 27
Herroelen, Willie, 156
Holt, Kat, 194
How to Manage in a Flat World (Bloch/Whiteley), 136
Huether, Derek, 173–174
Huntercombe Group, 194

Inglis, Graham, 18, 66
invoices, 11, 36
IS Project Management Handbook, 4
issues
    logs, 83, 205–207
    management, 83–85

James, Dr Ady, 77–9, 83–84
Jordan, Andy, 183
Juli, Thomas, 133

Kennedy, John F., 146
Kirby, Liz, 93–94
Kolb, Judith, 189
Koppelman, J. M., 19–20

Lake Forest Graduate School of Management, Chicago, 184
Le Guin, Ursula K., 63–64
leadership, 133–4
Leadership & Organization Development Journal, 189
Leadership Principles for Project Success (Juli), 133
Leading on the Edge International, 12
Leading Successful PMOs (Taylor), 131
Leus, Roel, 156
LinkedIn, 177–178, 202
Lissak, Robin, 142
lists, 171–172
Lloyds TSB, 55, 58
Lost Lovers' Ball, 146
Love, P. E. D., 81

McDonald, Peter, 14–15
McDowall-Long, Kimberly, 192
Madsen, Susanne, 105–106, 120–121
Major Projects Report (MOD), 74
Makar, Dr. Andrew, 177–178
management
    and leadership, 134
    matrix, 135–137
management reserves *see* budgets, contingency fund
Marshall, Alison, 7–8, 181–182
meetings, 8, 25, 79, 122, 137, 158–159, 188–189
mentoring, 191–193
Met Office, 45
Microsoft SharePoint, 68
milestones, 162
Ministry of Defence, 74
Montague, Celine, 175
Mullaly, Mark, 62
Mullard Space Science Laboratory, 83
multi-tasking, 166–167
Murch, Richard, 186

network-based techniques, 150–151
nuclear waste, 4

Obeng, Eddie, 95, 117
office politics, 181–183
OGC P3O standard, 49

Pacelli, Lonnie, 209
parking, 7–8
Particle Physics & Astrophysics Research Council (PARC), 83
parties, 128–129
Patient Record Project, 111
peer reviews, 28–31
Peplow, Phil, 97–98
Perfect Projects (Obeng), 95, 117
photography, 127
piloting, 47
Pinterest, 178
planning, 161
PMI Agile Community of Practice, 173
PMOmid (PMO knowledge pyramid model), 131
portfolio management, 167
PricewaterhouseCoopers, 142
Project Charter/Initiation Documents, 106
project governance models, 122
project management
    budgets, 12, 27
    documentation, 194–197
    engaging the audience, 125–127
    facilitation, 187–190
    foreseeing situations, 209
    golden triangle, 43, 56
    handover, 97–99
    managing yourself, 169, 171–172
    mentoring, 191–193
    network-based techniques, 150–151
    office politics, 181–183
    parties, 128–129
    personal brand, 177–180
    records, 175–176
    sharing knowledge, 202
    soft skills, 113–115
    sponsors, 6, 8, 19, 118–119, 198–199, 200

219

stakeholders, 120–124
success criteria, 61
teams *see* teams
timesheets, 11, 35–37
'walking around management', 78
Project Management Answer Book, The (Furman), 56
Project Management: Best Practices for IT Professionals (Murch), 186
Project Management Coaching Workbook, The (Madsen), 105, 120
Project Management Demystified (Reiss), 154
Project Management for Dummies, 13
Project Management Interview Questions Made Easy (Makar), 177
Project Management Journal 2010, 13
Project Management Jump Start (Heldman), 27
Project Management Offices (PMO)
    archiving, 200
    budgets, 24
    document assumptions, 87–88
    peer review, 30
    storing information, 175
    timesheets, 35
    tracking benefits, 59
    usage of, 130–132
    wikis, 71
    zombie projects, 173
Project Management for Small Projects (Rowe), 166
project management software, 135
Project Management Specialist Group (BCS), 209
Project Pain Reliever (Jordan), 183
Project Pain Reliever (Wood), 54
Project Perfect, 86, 116
Project Whirlwind, 16
projects
    Agile, 64, 101, 141, 150, 158, 174
    archiving, 200–201
    assumptions, 8
    baseline schedules, 155–157
    benefits, 40–42
    calculating estimate to complete, 12

change, 51–54, 96
contingency budgets, 18–19
costs of, 9
delegation, 160–162
dependencies, 162, 163–164
documentation, 194–197
failing, 198–199
fitting, 48–50
life cycle, 9–10
management of, 32–36
monitoring, 55
multiple, 165–168
Non-Agile, 64
planning, 139, 141–142, 143–145
portfolios, 50
post-project reviews (PPR), 70–73, 157
recording, 35
recovery exercises, 134
requirements, 100–101
resources required, 4
segmentation, 151
showstoppers, 184–186
signing off, 111–112
size, 45–47
small, 149–151, 165–166
time-bound, 146–148
tracking benefits, 58–62
tracking budgets, 11, 23–24, 26–27, 30
zombie, 173–174
PROMS-G, 209–210
publication, 143
Pullen, Ruth, 210
PUST project, 141–142, 158

quality
    activities, 57
    assurance, 56–57
    control, 56
    plans, 55–57
    standards, 56–57

records, 175–176
Reiss, Geoff, 154
Rescue the Problem Project (Williams), 31
risks
    contingency funds, 18
    identification, 74–76, 209

logs, 203–234
management, 75, 77–82
operational, 77
programmatic, 77
sub-system level, 78
system level, 78
robots, 91
Rodgers, Bob, 28–29
Roffey Park Institute, 21–23
Rothwell, William, 189
Rowe, Sandra, 166

St George's Park, 89
salary costs, 24
Salm, G., 81
Sampson, Marina (now Tumblety), 191–192
schedules, 144
Schmaltz, David, 155–156
Schulte, Ruthanne, 20
scope
    definition, 89–93
    documenting, 93–94
    golden triangle, 43
    human resources project, 55
    post-project review, 70–73
    projects, 117–118
    stakeholders, 133
    statements, 90
Second Order Project Management (Cavanagh), 9
Secretary of Veterans Affairs (US), 112
Shim, J. K., 8–9
Shrub, Avraham, 151
Siegel, J. G., 8–9
signing authority, 26–27
Simpson, Claire, 160
slide:ology:Art and Science of Creating Great Presentations (Duarte), 127
Smallman, C., 75
Smith, Mark, 89
Songhurst, Caroline, 152
Spanitz, Eric, 184–186
SPARKS programme, 125–126
sponsors, 6, 8, 19, 118–119, 198–199, 200
spreadsheets, 5
stakeholders
    ambiguity, 64
    analysis of, 120–124

benefits, 58–59, 62
brainstorming, 4
change, 53
communication, 93
documentation, 194
estimating, 5
failing projects, 198
INFORM, 124
and leadership, 133
management, 123–124
managing plans, 139
mini reviews, 70
momentum, 142
office politics, 181, 182
one-document approach, 135–136
piloting, 47
project planning, 139
project teams, 103
quality, 56
requirements, 101
scope, 89–90, 133
scrapping projects, 198
vision, 106–7
standardised reporting, 136
statements, 90
Steering the Craft (Le Guin), 63
strategy triangles, 49
Support Vehicle project, 74
Swanson, Monica, 135–136
Sweeney Communications, 125
Sweeney, Jo Ann, 125–126
Synergest, 184

Taming Change with Portfolio Management (Durbin/Doerscher), 50
Taylor, Peter, 131
teams
  collaboration skills, 136–137
  culture, 108–110
  introduction, 103
  multi-cultural, 108–109
  performance, 114
  planning, 143–144
  showstoppers, 186
  vision, 105–107
  zombie projects, 174
Ten Six Consulting, 28
Thomas, Janice, 62
time recording, 37
timesheets, 11, 35–37
Tinker, Antony, 48–49
Total Quality Management (TQM), 188
training
  courses, 8, 38–39
  sponsors, 116–119
Trompenaars, Fons, 109
True North (management consultancy), 155
Truss, Catherine, 94
trust, 95
Turbit, Neville, 86–87, 116–117
Turner & Townsend, 89
Twitter, 177–178, 202

Unachukwu-Hamori, Adrienn, 108–109
United States Civil Engineering Research Foundation, 157
Upstairs at Duroc (journal), 143–144

VA/DOD Health Executive Council, 112
Valentine's Day, 146
van Maurik, John, 189
version control, 66–69
video, 127
Viles, Gemma, 35–36, 130

webinars, 38–39
What Business Really Wants from IT (White), 60
White, Terry, 60
Whiteley, Philip, 136
WICE, 188
wikis, 71, 98
Williams, Todd, 31
Wood, Michael, 54
Woolliams, Peter, 109
workstream leaders, 160–162
WOW project, 45–46
Wright, J. Nevan, 24
Wrike, 194

X-rays, 40

Zombie Project Management (Huether), 174
zombie projects, 173–174